CUTTING AND GRINDING FLUIDS: SELECTION AND APPLICATION

Jeffrey D. Silliman
Editor

Richard Perich
Staff Editor

Published by

Society of Manufacturing Engineers
Reference Publications Division
One SME Drive
P.O. Box 930
Dearborn, Michigan 48121

CUTTING AND GRINDING FLUIDS: SELECTION AND APPLICATION

Copyright © 1992
Society of Manufacturing Engineers
Dearborn, Michigan 48121

Second Edition
First Printing

All rights reserved including those of translation. This book, or parts thereof, may not be reproduced in any form or by any means including photocopying, recording, or microfilming or by any information storage and retrieval system, without permission in writing of the copyright owners. No liability is assumed by the publisher with respect to the use of information contained herein. While every precaution has been taken in the preparation of this book, the publisher assumes no responsibility for error of omission. Publication of any data in this book does not constitute a recommendation of any patent or proprietary right that may be involved or provide an endorsement of products or services discussed in this book.

Library of Congress Catalog Number: 92-809773
International Standard Book Number: 0-87263-423-X
Manufacturing in the United States of America

PREFACE

The American Society of Tool and Manufacturing Engineers (ASTME) first published "Cutting and Grinding Fluids: Selection and Application" in 1967. It was written to consolidate the "state-of-the-art" knowledge at that time. In the intervening 25 years ASTME has become SME (Society of Manufacturing Engineers) and numerous significant advances in the "state-of-the-art" have occurred. This revision has sought to capture these advances in pursuit of the book's original purpose of providing comprehensive, accurate information required by today's engineers and managers in selecting and managing cutting and grinding fluids.

There have not only been major advances in metal cutting technologies such as creep-feed grinding, but also changes in both operator and environmental requirements that have added important considerations to the selection, management and disposal of these fluid products. Our engineers and managers must have complete, up-to-date knowledge of the available fluid types, functions, application methods, quality controls and operator and environmental considerations to properly compete in a worldwide manufacturing arena that requires maximum productivity and minimum waste.

This book was updated through the sponsorship of Clyde A. Sluhan, one of the pioneers of the science of metal cutting and founder of Master Chemical Corporation. Mr. Sluhan is a SME Fellow and was chairman of the ASTME Metalworking Fluids subdivision that originally authored this work. Technical review and updates were provided by numerous experts. Special thanks are due to Dr. Stewart Salmon of Advanced Manufacturing Science & Technology who contributed the chapter on creep-feed grinding and superabrasives and Mr. Cleve Patten of Master Chemical Corporation who expertly and patiently layed out and typeset this revision through many drafts.

It is our hope that this book continues to represent the comprehensive treatment of cutting and grinding fluids and their selection, application and management. We trust that this body of information will continue to aid the user in developing better manufacturing methods and products for their operations and our overall metalworking technology.

Jeffrey D. Silliman

Toledo, Ohio
October, 1992

TABLE OF CONTENTS

Nomenclature		1
CHAPTER 1	HOW FLUIDS FUNCTION	5
	Basic Considerations	5
	Analysis of Cutting Fluid Action	12
	The Cooling Mechanism	19
	The Lubrication Mechanism	26
	The Shear-Strength-Reduction Mechanism	28
	Interrelationship of Mechanisms	29
CHAPTER 2	FLUID TYPES	35
	Cutting Oils	36
	Emulsified Oils (Water Miscible)	41
	Chemical and Semichemical Fluids (Water Miscible)	43
	Gaseous Fluids	46
	Miscellaneous Fluids	47
CHAPTER 3	HOW CUTTING FLUIDS ARE APPLIED	49
	Manual Application	50
	Flood Application	52
	Air-Carried Mist	56
	Chilled Fluid Application	59
CHAPTER 4	SELECTING FLUIDS FOR MACHINING AND GRINDING PROCESSES	63
	Turning	76
	Boring	76
	Facing	77
	Grooving	77
	Forming	78
	Cutoff	78

	Box Turn	78
	Trepanning	79
	Milling	79
	Drilling	84
	Gundrilling	85
	Counterboring	85
	Spot Facing	86
	Countersinking	86
	Reaming	86
	Broaching	87
	Tapping (Internal Thread Cutting)	87
	Single Point Thread Cutting	87
	Chasing (External Thread Cutting)	87
	Grinding	88
	Gear Hobbing	92
	Gear Cutting	92
	Gear Shaping	93
	Gear Shaving	93
	Sawing	94
	Abrasive Cutoff	95
	Honing	95
	Lapping	96
CHAPTER 5	ACCEPTABILITY OF CUTTING FLUIDS	97
	Physical Laboratory Tests	99
	Chemical Laboratory Tests	101
	Mechanical Laboratory Tests	105
	Metallurgical and Chemical Compatibility	107
	Human Compatibility	111
CHAPTER 6	QUALITY CONTROL OF METAL CUTTING FLUIDS	119
	Handling Methods	119
	Water Sources and Composition	120
	Proper Mixing Procedures	124
	Bacterial Effects and Prevention	125
	Fungi	129
	Fluid Cleaning Methods	129
	Cleaning Machine Tools and Circulating Systems	132
	Disposal Of Water-Base Fluids	133
	Waste Minimization	134
CHAPTER 7	COMPATIBILITY OF LUBRICANTS WITH CUTTING/GRINDING FLUIDS	137
	Lubricant Systems	137
	Total Loss Lube Systems	141
	Multipurpose Machine Tool Lubricants	141

CHAPTER 8	ECONOMIC CONSIDERATIONS IN CHOOSING A FLUID 143
	Economic Justification 145
	Overall Results 155
CHAPTER 9	TROUBLESHOOTING CUTTING AND GRINDING FLUID APPLICATIONS 161
	Machining Operations 161
	Grinding Operations 165
	Symptoms, Causes, and Correction Of Problems 167
CHAPTER 10	CREEP-FEED GRINDING AND GRINDING WITH SUPERABRASIVES 181
	Creep-Feed Grinding 181
	Grinding With Superabrasives 184
CHAPTER 11	HEALTH AND SAFETY, ENVIRONMENTAL AND REGULATORY CONSIDERATIONS OF METALWORKING FLUIDS 187
	Health and Safety 187
	Environmental Considerations 189
	Regulatory Considerations 190

APPENDIX 193

GLOSSARY 199

BIBLIOGRAPHY 203

INDEX 209

LIST OF TABLES

Table 1-1	Density, Volumetric Specific Heat, Thermal Conductivity and Viscosity of Selected Fluids	15
Table 1-2	Effect of Various Cutting Fluids in Lowering the Tool–Workpiece Temperatures When Machining Steels	22
Table 3-1	Fluids for Use at Subambient Temperatures	60
Table 4-1	Recommendations for Cutting and Grinding Fluids and Application Methods	64
Table 4-2	Cutting Grinding Fluid Codes	74
Table 4-3	Application Methods	75
Table 5-1	Ratings of Degree of Separation or Turbidity	100
Table 8-1	Elements of Total Manufacturing Costs	156
Table 8-2	Cutting and Grinding Fluid Justification-Cost Work Sheet	158
Table 8-3	Summary Of Production Cost As Cutting Fluid Application Is Investigated	159
Table 9-1	Machining Operation Problems	170
Table 9-2	Grinding Operation Problems	172
Table 9-3	Maintenance Problems	173
Table 9-4	Physiological Problems	176
Table 9-5	Economic Problems	177
Table 11-1	Haulaway Cost For Spent Coolant By Region	190

NOMENCLATURE

Over the years, jargon becomes part of any technology and various names are given to items and systems peculiar to that technology. Many different names for the same thing are correct, but interchangeable usage is confusing. Metalworking fluids are no exception; a single type of fluid is called by many names. To clarify this and make this book more understandable to readers, the following nomenclature will be used throughout:

A. Chemical Fluids (sometimes called "Synthetic Fluids").

1. *True solutions*-clear (transparent) but possibly colored, consisting of inorganic and/or organic materials dissolved in water as shown in Figure A. The dissolved materials may be considered to be randomly dispersed within the solvent water molecules. Usually, the solution's surface tension is not much lower than that of plain water. The primary functions of true solution fluids are to cool and to inhibit rust. They usually have very low lubricating values.

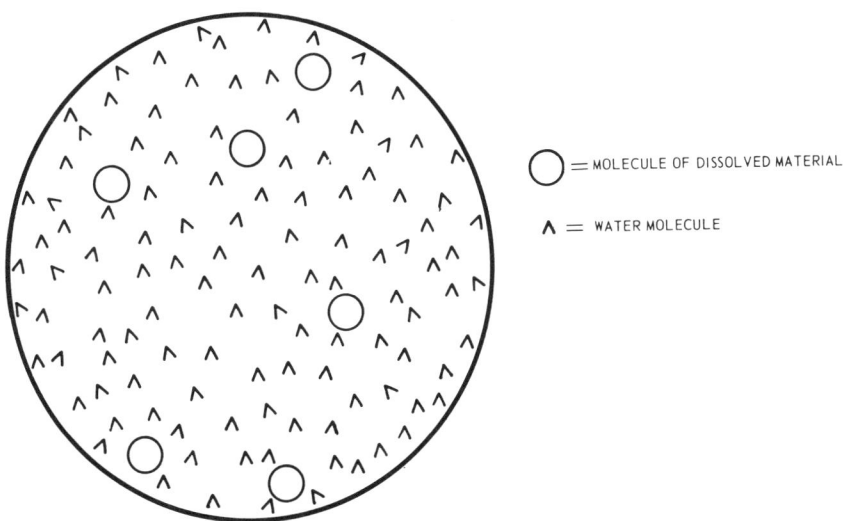

Figure A. *Pictorial representation of true solutions.*

2. *Surface-active*-water solutions of type (1) above to which anionic and nonionic surface active agents have been added. These lower the surface tension of water and form colloidal aggregates (micelles) among the surface-active molecules as shown in Figure B. These fluids are quite clear by casual inspection and have good lubricating qualities.

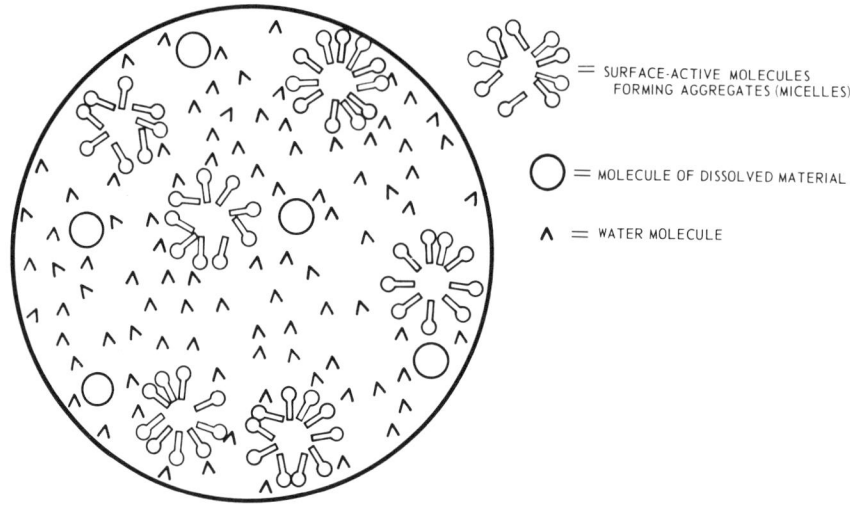

Figure B. *Pictorial representation of surface-active type.*

Improved lubricating or extreme pressure (EP) qualities are provided by incorporating sulfur, chlorine, or phosphorus additives.

B. Emulsions (commonly called "Water Soluble Oils," "Water Emulsifiable Oils" or "Emulsifiable Cutting Fluids"). These fluids are a suspension of oil droplets (mineral, paraffinic, or naphthenic base oils) in water. They are made by blending the oil with emulsifying agents and other materials so that oil droplets of 0.0002 to 0.00008 in. in diameter form when mixed with water as illustrated in Figure C. The addition of animal or vegetable fats or oils or other esters produces "super-fatted" emulsions of greater lubricating value. The addition of sulfur, chlorine, or phosphorus products produces fluids of even greater lubricating value. These are often designated as extreme pressure (EP) emulsions.

C. Semichemical Fluids (sometimes called "Semisynthetic Fluids"). These fluids are essentially a combination of types (A) and (B), but with the following additional characteristics:

1. Lower oil content (e.g., 5 to 45%) than type (B) fluids.
2. Higher content of emulsifying or surface-active molecules plus "blending agents" than type (B), resulting in smaller average oil droplet diameters as shown in Figure D. These materials can be plain, "super-fatted," or extreme pressure (EP) types.

D. Cutting Oil. A cutting fluid which may be an oil of petroleum, animal, or vegetable origin, either singly or in combinations. The petroleum oil may vary in source such as naphthenic or paraffinic, and in range of viscosity from very low to very high, depending upon the intended application.

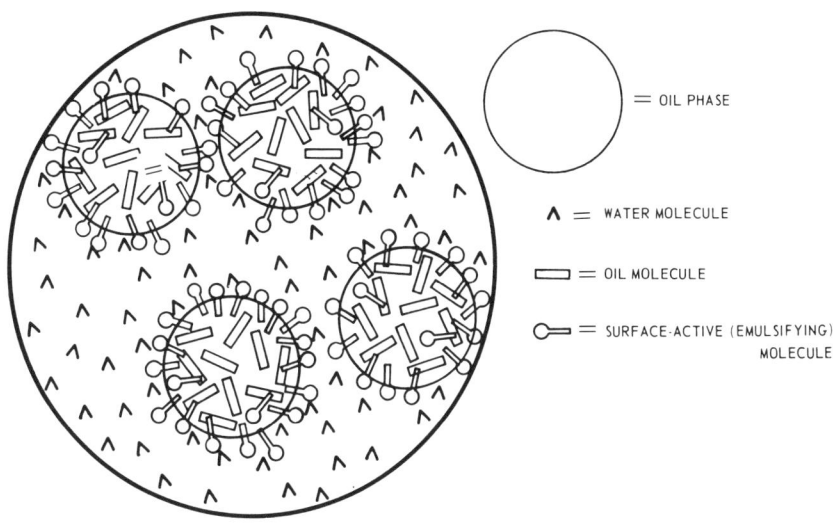

Figure C. *Pictorial representation of emulsions.*

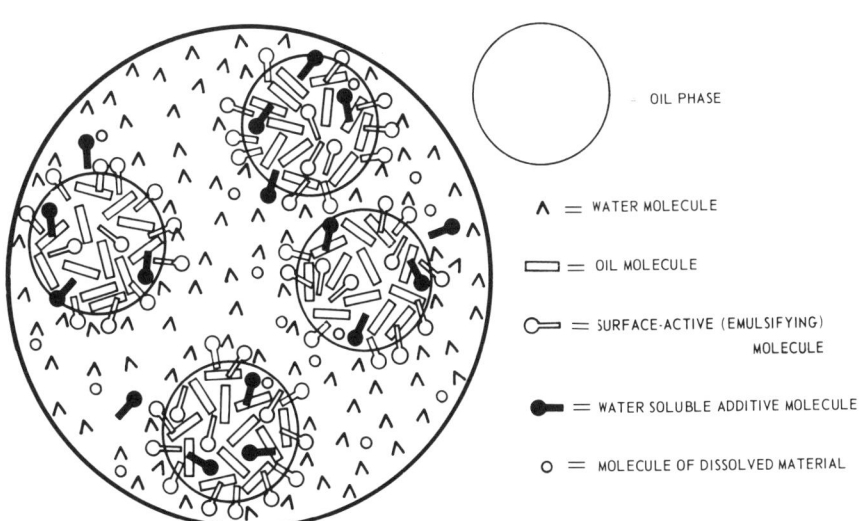

Figure D. *Pictorial representation of semichemical fluids.*

The addition of animal or vegetable oils (fatty oils) increases the wetting action of the petroleum oil blend and improves the lubricating qualities, particularly at higher temperatures.

Chlorine, sulfur, or phosphorus additive agents can be incorporated, as depicted in Figure E, to improve lubrication effects at even higher temperatures and pressures than possible with petroleum oils alone or combinations of petroleum oils and fatty oils.

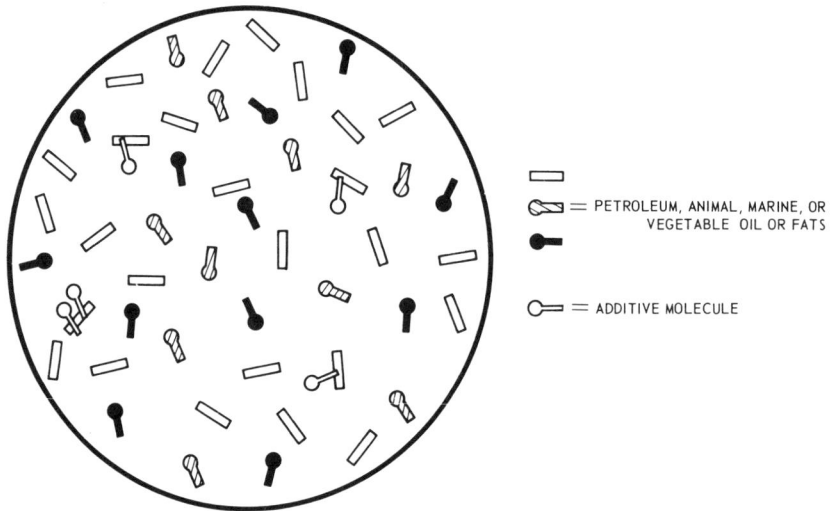

Figure E. *Pictorial representation of cutting oils.*

CHAPTER 1
HOW FLUIDS FUNCTION

Metal cutting, practiced in some form since ancient times, assumed great importance during the last 200 years as machining played a major role in the Industrial Revolution. The use of cutting fluids came instinctively to early machinists, who applied lubricants initially "by brushing them on" and later by flooding the cutting area to reduce friction and heat generation.

The first publication on cutting fluid applications appears to be a book by Northcott who, in 1868, reported that lathe productivity could be materially increased by this means (1). Since that time, the economic advantages of an increasing variety of cutting fluids have grown steadily.

Although volume, types, and applications of cutting and grinding fluids have kept pace with industrial development, an understanding of how they work has followed very slowly, particularly in the area of apparent lubrication effects in low-speed cutting. An early worker in 1881, Mallock, sensed part of a perennial mystery when he wrote, "Lubricants seem to act by lessening the friction between the face of the tool and the shaving, it is difficult to see how it gets there. Perhaps into the substance of the shaving"(2). Only by intensive research during the last 45 years have this and related questions been given reasonably satisfactory answers; yet other difficulties remain unsolved.

Detailed discussions of the application, composition, and selection of cutting and grinding fluids appear in Chapters 2, 3, and 4. This chapter is limited to a discussion of the "why" and "how" of the cutting fluid's basic role. The differential significance of several possible mechanisms of fluid action is examined under various machining conditions. An interpretation is made suggesting that each of three major mechanisms are valid and interdependent. Their relative importance differs with various fluids and with changes in application technique.

BASIC CONSIDERATIONS

The two major effects of cutting and grinding fluids - cooling and lubrication - are best accomplished by water-based and oil-based fluids, respectively. But it has become apparent that both coolant and "lubricant" action result (to different extents) from the use of both classes of fluids, with more functional overlap than has been generally realized.

It is important to consider at this point that, while lubricant type fluids do seem to reduce cutting friction, it is a cutting-force ratio involving shear that is being measured - not a coefficient of ordinary sliding friction. Higher temperatures normally facilitate the shear flow of metals; lubricants, *per se*, have not been understood to have a direct effect on such processes. It is surprising, therefore, to find that both coolants and "lubricants" are capable of reducing cutting forces. Furthermore, by removing heat or preventing its generation, respectively, both manage to reduce cutting temperatures, extend tool life, and improve workpiece finish.

These apparent contradictions have increasingly tantalized persons involved with production, as well as laboratory cutting and grinding, during recent years. It is certain that the effects do not arise from specialized cutting geometries. In view of this, the subsequent discussion (wherever possible) is deliberately limited to mechanisms as they concern orthogonal or two-dimensional cutting. This simplification is intended to help clarify the important relationships which are complex enough without introducing unnecessary geometrical parameters.

Cutting and Grinding Geometries

The infinite variety of geometrical relationships possible between tool and workpiece defies brief representation except in terms of a simple, idealized model. Therefore, the two-dimensional diagram shown in Figure 1-1 has been chosen to represent schematically a cross-sectional view of orthogonal cutting. Here the cutting edge of a wedge-shaped tool is set perpendicular to the line of relative motion of tool and workpiece.

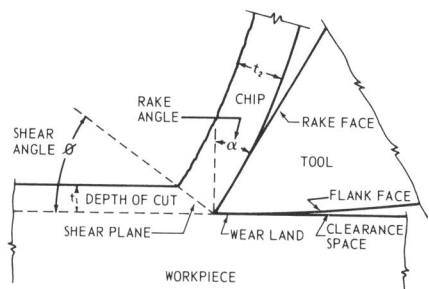

Figure 1-1. *Two-dimensional metal cutting diagram.*

In Figure 1-1, chip thickness, t_2 and depth of cut, t_1 bear a simple geometrical relationship to each other and to the shear angle, ϕ. This provides a meaningful parameter: a cutting ratio (t_1/t_2) which is also called the "chip shortening ratio" because it is numerically equal to the length of the chip divided by tool travel.

$$\text{Cutting Ratio } r_c = t_1/t_2 = \sin \phi \qquad (1\text{-}1)$$

BASIC CONSIDERATIONS

Merchant (3, 4) has presented a useful "slip-line-theory" analysis of orthogonal cutting in which the line of the shear angle, ø, is understood to denote the locus of a "shear plane."

In this simplification, shear has been considered to be localized in the "shear plane." Shear stress actually results in strain over a larger region; this is called the "primary deformation zone" in Figure 1-2. The secondary zone of deformation indicated in Figure 1-2 represents the built-up edge region, where, dependent upon cutting speed, strain-hardened workpiece material tends to "stagnate" into a so-called secondary tool tip.

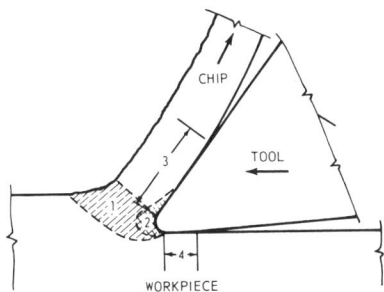

1) Primary deformation zone (shearing, strain hardening)
2) Secondary deformation zone (built-up edge)
3) Primary friction zone (tool/chip interface)
4) Secondary friction zone (tool/workpiece interface)

Figure 1-2. *Presumed zones of shear and friction in metal cutting* (5).

The rake angle, α, obviously a critical machining variable, is the angle at which the chip must slide up the rake face of the tool and affects the forces of sliding friction in the tool/chip interface, the primary friction zone of Figure 1-2. Merchant's slip-line calculations allow the writing of simple relationships between the work of cutting, W_s, and shear strength of the metal, S_s, involving only α and the cutting ratio, r_c. The amount of heat produced when α is constant is inversely proportional to the shear angle, ø. This angle, in turn, is inversely proportional to the coefficient of friction, μ, between tool and chip. Another zone of friction shown in Figure 1-2, is the tool/workpiece interface, a region of wear and sliding friction. While normally of secondary significance, this depends on the extent of tool wear, tool material, and other factors.

Function of Cutting Fluids

Most metal cutting processes which result in a chip need cutting fluid action. Very little (1 to 3%) of the work of metal cutting is stored as residual stresses in the workpiece or chip, more than 97% appearing as heat. Of this, about 2/3 is expended in the sticking friction in the shearing zone, and 1/3 in the sliding friction at the tool/chip, tool/flank interfaces at normal chip thicknesses (6).

Aqueous fluids, if they served only as coolants, would exert no influence on heat generation in either zone, but simply remove heat and reduce catastrophic tool failure with the possibility of actually increasing abrasive tool wear. Oil-based fluids, considered only as lubricants, would reduce heat generation (plus give minor heat-removal effects), but almost exclusively in the zone of sliding friction, not in the shear zone.

As developed in subsequent sections, some of the results observed with the two broad classes of fluids bear out these expectations. Other results, particularly those concerned with effects on heat generation during shearing, have suggested a third mechanism involving alteration of the microstructure of metals during deformation prior to shearing.

Cutting at Low Speeds. Good surface finish is difficult to attain in dry machining of the common metals at cutting speeds below about 200 fpm (61 m/min) (7). This fact, other than practical production-rate considerations, is the outstanding fault of low-speed metal cutting. The effect appears to stem from kinetic considerations concerning the widening of the secondary deformation zone of Figure 1-2 (8).

Ernst and Merchant (9) showed that the resulting built-up edge in the tool was the major cause of roughness at low cutting speeds. While this edge material protects the tool tip from abrasive wear, it sloughs off continuously in the form of workhardened fragments (10). Furthermore, since the widened zone of deformation involves the workpiece itself (11), it is characteristic for the work surface to register overall work-hardening --but with discontinuities which evidence gouging.

In addition, at slow speeds, the vibrational harmonics of the tool/workpiece setup seem to be involved with the slip-stick of strain-hardened discontinuities. This factor would be eliminated by kinetic factors at higher speeds, just as it is relieved by the chemical lubricant action of cutting fluids at low speeds.

Cutting at Higher Speeds. As cutting speeds rise above 200 fpm (61 m/min), the deformation zones are reduced, and the built-up edge tends to disappear. Since nearly as much useful work is done, heat generation is concentrated in a smaller region near the cutting edge (12). This means a concentration of thermal effects producing higher cutting temperatures, resulting in a more nearly plastic deformation of the softened workpiece material. If tool wear were not affected negatively, and if the tendency for the hot chip to weld to the tool were not a factor, speed could be beneficial. Trigger (13) demonstrated that cutting temperatures with NE 9445 steels show a logarithmic increase at speeds rising from 50 to 500 fpm (15 to 152 m/min). At the same time, at a given depth of cut, cutting forces were decreased by approximately 25% and chip hardness was doubled (depending on the original hardness of the workpiece).

At higher speeds, the chips, besides being thinner, are smoother and softer, reflecting the general improvement in finish of the workpiece. But since the chips at a given cut are thinner, the cutting ratio (r_c) is smaller. Thus, cutting forces are lower so that appreciably less power is needed per unit of metal removed under these conditions (14).

Unfortunately, the hotter and softer chip conveys its thermal energy to the rake face of the tool where crater wear adds to the expected increase in wear of the hotter, and now unprotected, tool tip. The result is an exponential decrease in tool life which becomes catastrophic at excessive speeds. Furthermore, the higher temperatures of both chip and tool produce a related problem: the seizure and incipient welding of these two structures. Thus, coolant action is required to curb high cutting temperatures, and "lubricant" action is needed to reduce friction (or its equivalent) in a region further up the chip (i.e., the tool/chip interface) for practical machining.

Cutting Hard and Brittle Metals. Brittle metals consist of aggregations of large crystals with active cleavage-plane discontinuities. The crystals may be hard but, in cutting, fracture at the discontinuities is easy so that built-up edge, temperature rise, and cutting forces are minimized; consequently, dry cutting is both natural and practical. Other metallurgical characteristics remaining constant, increases in ductility result in more plastic flow, with increase in built-up edge, formation of the continuous chip, temperature increase, and larger cutting forces. Here, coolant and lubricating fluids become important.

In more homogeneous structures, increases in hardness resulting from variations in annealing give a different picture. Trigger (13) demonstrated that small but definite increases in cutting temperature were experienced with increase in hardness of steel at cutting speeds from 50 to 300 fpm (15 to 91 m/min). In the same study, particularly at 100 to 300 fpm (30.5 to 91 m/min), substantial decreases in chip thickness were observed as hardness increased, this effect being very speed-dependent at lower hardness levels. A similar effect was seen in terms of increased chip hardness.

Figure 1-3 shows the interrelationship between unit pressure, temperature, and time when machining the more difficult-to-machine (high-strength) metals. Speeds must be kept low to prevent excessive heat buildup in the tool. Any chemical lubricating effects from a cutting fluid will be more pronounced because of the lower speeds necessary.

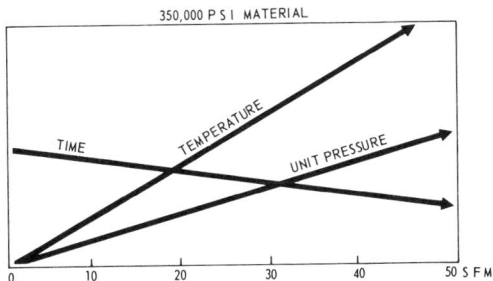

Figure 1-3. *Interrelationship between unit pressure, temperature, and time when machining the more difficult-to-machine metals. (Courtesy, Master Chemical Corporation)*

Figure 1-4 shows a simple representation of the interrelationship between unit pressure, temperature, and time for possible chemical reactions

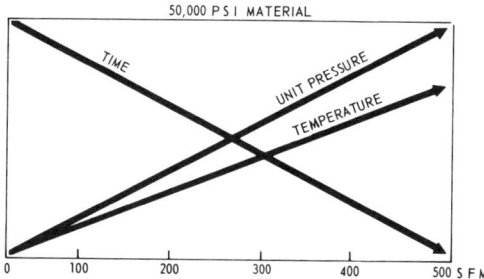

Figure 1-4. *Interrelationship between unit pressure, temperature, and time for possible chemical reactions with cutting fluids when machining low psi. metals. (Courtesy, Master Chemical Corporation)*

with cutting fluids when machining easy-to-machine (low-strength) metals. Higher possible speeds result in higher temperatures and higher unit pressures between chip and tool (due to decreased chip thickness). However, counteracting these two phenomena, which would tend to increase the rate of possible chemical reactions between the cutting fluid and the cutting zone, is the decreased time possible because of higher speed of chip travel. Hence, cooling becomes more important than lubrication in cutting at higher speeds.

Hardness itself, therefore, does not present difficulty when choosing a cutting fluid. The "benefits" of cutting metals of increased hardness are purchased at the price of higher cutting temperatures and greater tool wear, just as the machining of homogeneous metals at lower hardness levels.

Function of Grinding Fluids

In the previous sections, the basis for the action of different kinds of cutting fluids has been examined. An essential factor in this study has been the use of simplified geometries as in a two-dimensional view of orthogonal metal cutting. Grinding geometries, characterized by a variety of complex negative rake tool surfaces, do not lend themselves to analysis of this sort. This is especially true since slip-line theory analysis of cutting forces has not been able to explain the dynamics of negative rake cutting. According to Shaw, grinding chip formation is more nearly analogous to the extrusion process with a plasticized chip extruded through an unsupported region in the metal surface immediately ahead of the grinding grit.

The basic functions and requirements of grinding fluids are similar to those of cutting fluids, despite pronounced differences between the geometry and dynamics of the processes. Beyond the rake angle and random orientation of cutting faces, the temperatures involved in grinding are higher (2000° to 3000°F [1090° to 1650°C]) as are grinding speeds which are in the 2000 to 16000 fpm (610 to 4900 m/min) range (15).

In cutting, tool wear is normally undesirable, but in grinding, some features of wear are desirable. Ordinary wearing away of the abrasive grains is undesirable because flat areas develop which generate excessive heat. Wear by grain fracture is needed, however, to replenish sharp cutting edges. Bond fracture is also important, so that a worn grain is expelled (16, 17, 18).

Most of the heat generated in grinding with conventional abrasives is conducted into the workpiece; less is removed by the fine curly chips and by the wheel. Even a good coolant can influence this transfer only slightly. The cooling effectiveness of a fluid can be improved greatly by forcing it into the cutting zone as a high velocity jet (15), and by giving full play to its wetting characteristics (16, 17, 18). Conversely, superabrasives (diamond and cubic boron nitride) do have significant heat transfer capabilities and conduct a significant percentage of the heat generated into the wheel.

Heat generation and heat removal are somewhat more clearly differentiated in grinding than in cutting. Grinding oils with active chlorine and sulfur are effective in reducing heat *generation* and in providing good finish. Their cooling ability, however, is relatively poor. They are not useful when a high level of heat generation is inevitable, due to workpiece toughness and/or high linear speeds, just as in cutting operations.

The cooling function, as important as it is in practical grinding, does not register large changes in wheel wear and has little influence on heat generation. Dry grinding does not produce appreciably larger instantaneous local temperatures than wet grinding. The primary benefit of coolant use in grinding is the prevention of gradual workpiece temperature rise.

Chemically active fluids, however, reduce cutting forces, provide better surface finish, and increase wheel life. Wheel life is measured by an index called "grinding ratio," which is defined as the volume of workpiece material removed per unit volume of wheel wear.

Increases in the grinding ratio are dependent on the concentration of chlorine sulfur compounds in test fluids, as shown by Tarasov (16, 17, 18), which demonstrate their chemical effects. Stewart and Soderstrom (19), in comparisons of a large number of grinding oils, also showed that a good correlation exists between grinding ratio, normal and tangential grinding forces, and residual stress. Sulfur, as might be expected at the temperatures involved, was more effective than chlorine in these tests.

Thus, in grinding, the functions of fluids are primarily as chemical agents capable of reducing grinding forces and improving finish. The cooling function must be considered as secondary, though it may be essential to the success of a specific grinding operation.

ANALYSIS OF CUTTING FLUID ACTION

Although practiced and reported much earlier (1, 2), the routine application of cutting fluids to machining processes seems to date from the work of Taylor (20). He effectively demonstrated their very practical effect of prolonging tool life and/or permitting increases in rate or quantity of metal removal. As the art, and gradually the science, of cutting fluid technology grew, the importance and extent of concurrent benefits were realized: cutting force requirements were appreciably lowered, better finish was achieved, and the exact product size was more easily reached in production due to improved dimensional stability of the workpiece. Minor but important secondary benefits resulted: chips were flushed away from the cutting zone, machine-tool slideways were lubricated, the product was protected from corrosion, and product temperature was lowered, allowing easier handling and inspection.

Some problems and limitations were inevitable. Use of oily and aqueous fluids presented fire hazards, physical instability, and bacterial spoilage difficulties; the corrosivity or toxicity of some otherwise excellent fluids have countermanded their use. But, more importantly, even the best fluids could not solve the problems of poor surface finish in low-speed cutting and drastic decreases in tool life at high speeds. The general problem of basic tool wear cannot be considered solved, despite much progress. Certain combinations of metallurgical properties result in "low machinability" characteristics which cutting fluids, even at ideal cutting speeds, have not been able to improve. The best extreme-pressure-type additive cannot prevent segments of the workpiece material from welding to the tool face under conditions required for optimum machining of certain metals.

Finally, a limitation on understanding has dogged both the analytical and practical mind in machining technology. This has the form of a controversial and contradictory overlap in the interpretation of the modes of cutting fluid action between the so-called cooling and lubricating mechanisms. Great progress has been made in recent years by researchers in this field. The results are beginning to substantiate old theoretical speculations about a third mechanism involving the environmental effect of chemically active fluids in altering the shear-flow characteristics of metal under shear stress. Future developments based on this new understanding may provide improved cutting fluids to solve both the old and new problems.

Physical Limitations

The term "fluid" refers to matter in either the liquid, critical, or gaseous states. The flow of liquids through microscopic orifices under a pressure gradient is seriously hampered by viscous resistance, the flow of gases less so by several orders of magnitude. Thus, gases can flow (or diffuse) more rapidly than liquids through small openings.

These points are significant. Clearances in metal cutting are small, particularly in the cutting zone itself. The ability of a fluid to gain access to the desired area is essential. Gases driven into the cutting zone at high speeds have shown remarkable efficiency in special applications (21, 22, 23). Liquids show some improvement in effectiveness when applied as a high-velocity stream (24).

Assuming access is achieved, however, liquids have two important advantages over gases. First, the higher viscosity of liquids promotes adhesion to the work and prevents them from being rapidly squeezed out of a confined space. Second, liquids are orders of magnitude better than gases in important thermal properties, including higher specific heats, heats of vaporization, and thermal conductivities. In this regard, water-based fluids (and dilute emulsions) are superior to oils. Water has a specific heat of 1.0 compared to 0.45 for a hydrocarbon oil. This means that, for a given amount of heat input, an equal weight of water compared to oil will have a resultant lower temperature as seen in Figure 1-5. Further, water will transfer heat from two to three times *faster* than oil. This is illustrated in Figure 1-6.

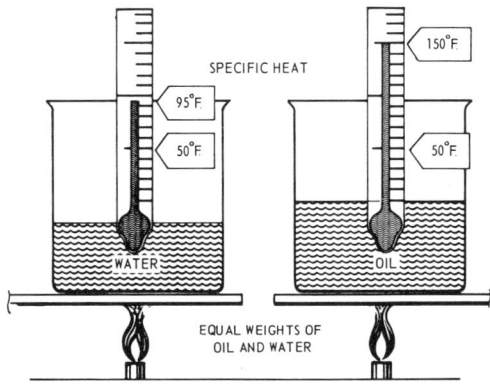

Figure 1-5. *Effect of same amount of heat on heat rise of equal weights of water and oil. (Courtesy, Master Chemical Corporation)*

To perform better than gases, however, liquids must make effective contact with metal surfaces. Such contact is limited by their wetting ability. A high contact angle at the metal/gas/liquid interface may cause liquids to "bead up" instead of spreading completely over the surface as depicted

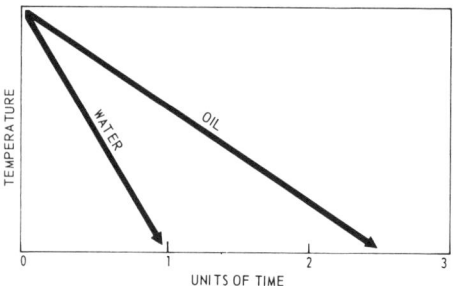

Figure 1-6. *Heat transfer rate of equal weights of oil and water. (Courtesy, Master Chemical Corporation)*

in Figure 1-7. When the interfacial tension between liquid and metal is zero, perfect "spreading" occurs and the energy of the system becomes lower. If the fluid contains chemically bound chlorine and/or sulfur, chemical reactions can occur with the metal surface, with a resultant further decrease in the energy of the liquid/metal system. In modern terms, such action would be described as involving the in-depth shear properties of the metal.

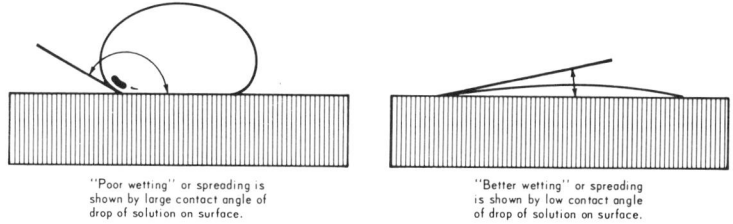

Figure 1-7. *Difference in wetting effect between two fluids of high and low surface tension. (Courtesy, Master Chemical Corporation)*

Thermal Properties

It has already been indicated that the static thermal constants for cutting fluids fall into the following order: aqueous liquids > organic liquids » gases. A comparison of thermal properties between water and selected organic liquids having thermal characteristics related to cutting oil components is presented in Table 1-1. For a fair comparison of potential thermal effects of a cold liquid lying on a hot surface (especially where gases are concerned) a volumetric comparison is needed. In Table 1-1, the specific gravities and volumetric specific heat values in BTU/cu. ft./°F are given for water, five organic liquids, and three gases which are used as cutting

Table 1-1. Density, Volumetric Specific Heat, Thermal Conductivity, and Viscosity of Selected Fluids

Fluid	Specific Gravity at 68°F.	Volumetric Sp. Heat at 68°F. BTU/(cu.ft.)(°F.)	Thermal Conductivity at Indicated Temperature (°F.) BTU/(sq.ft.)(hr.)(°F.)/(ft.)	Absolute Viscosity at 68°F., millipoise
Carbon tetrachloride (liq.)	1.59	19.8	0.107 at 32	10.0
Carbon tetrachloride (vap.)	—	—	0.0041 at 115	—
Ethanol (liq.)	0.79	28.1	0.105 at 68	12.2
Ethanol (vap.)	—	—	0.0089 at 68	—
Benzene (liq.)	0.89	21.8	0.092 at 86	6.5
Benzene (vap.)	—	—	0.010 at 212	—
n-Hexane (liq.)	0.66	24.7	0.08 at 86	3.2
n-Hexane (vap.)	—	—	0.008 at 68	—
n-Heptane (liq.)	0.68	20.8	0.081 at 86	4.1
n-Heptane (vap.)	—	—	0.01 at 212	—
Water (liq.)	1.00	62.4	0.342 at 68	10.0
Water (vap.)	—	—	0.013 at 212	—
Air (gas)	0.075	0.028	0.014 at 32	0.18
	—	—	0.018 at 212	—
Nitrogen (gas)	0.075	0.028	0.014 at 32	0.18
	—	—	0.017 at 212	—
Carbon dioxide (gas)	0.114	0.024	0.008 at 32	0.14
	—	—	0.012 at 212	—

fluids, along with their corresponding thermal conductivities and viscosity values.

Note that while the volumetric specific heats are lower by a factor of 10^4 for gases as compared to liquids, the thermal conductivity values differ generally by only a factor of 10^2. Part of this difference, a factor of 10, stems from lower viscosity; the remainder is accounted for by faster diffusion rates. In comparing thermal conductivities of vapors of the liquids with those of gases at a single temperature, e.g., 212°F (100°C), it will be noted that differences are small, with carbon tetrachloride values the lowest. Historically, metalworking researchers have used carbon tetrachloride (CCl4) as a cutting fluid under experimental laboratory conditions for a number of valid reasons. However, because of its extreme toxicity, carbon tetrachloride should never be used as a cutting fluid under shop conditions. If cutting zone temperatures actually convert liquids into gases before they complete their role in cutting, any cooling action subsequent to vaporization is achieved no better by liquid than by gaseous fluids (25).

Further comparisons between the values for liquids indicates the superiority of water in heat transfer, with other organic liquids far behind. Carbon tetrachloride, again, shows inferiority as a coolant, a point of particular interest since its action in reducing cutting forces is superior.

As shown by Pahlitzsch (26), the three gases listed in Table 1-1 are effective as coolants when applied at temperatures only 50°F (27.8°C) below ambient, or at linear rates of supply only one order of magnitude higher than normally feasible for liquid coolants. It would again appear that the overall heat transfer ability of gases under dynamic conditions is certainly much greater than expected from the low volumetric heat capacities, and considerably greater than suggested by heat conductivity comparisons.

This emphasizes the fact that static thermal properties of fluids, which do not include an accounting of access limitations or viscosity characteristics, cannot adequately reveal thermal differences under dynamic conditions. In laminar flow of viscous fluids, flow retardation of layers of liquids becomes progressively greater in layers nearer and nearer a solid surface, so that the rate of heat removal is a direct function of viscosity. Overall studies of dynamic heat transfer abilities which combine such factors properly are, therefore, of considerably more significance than static values. For this purpose, Hain (27) used a thin-walled stainless steel tube heated by its own electrical resistance. Cutting fluids entering at 90°F (32°C) were circulated through it, and cooling effects were assessed in terms of the resultant temperature-lowering profile of the tube as the fluid passed down its length. In these experiments, water was compared with various emulsions, regular cutting oils, and a petroleum naphtha.

Of particular interest were comparisons of a chemical surface active and a "soluble oil" concentrate, blended in various proportions with water. Typical results, shown in Figure 1-8, indicate that the chemical surface

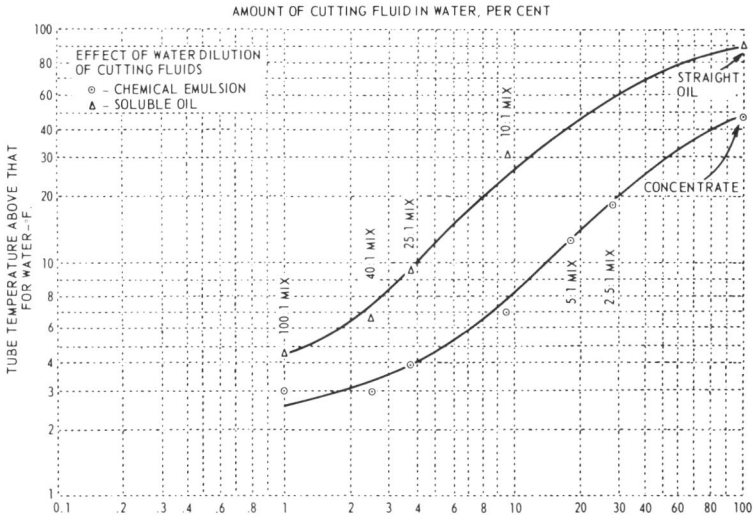

Figure 1-8. *Cooling ability (relative to water) of water-diluted cutting fluids of chemical emulsion and soluble oil types.*

active (at about 90% water) showed a regular increase in heat-transfer ability (relative to that of water) as dilution proceeded. The soluble oil gave a parallel improvement, but was less effective at all dilutions. Water was clearly the key ingredient. The importance of good water contact at the metal surface itself was also brought out in comparisons of two 10% emulsions of petroleum naphtha, one being oil external (W/O), the other water external (O/W). The calculated heat transfer coefficient of the latter was almost double that of the former.

Wetting and Penetration

If metals used in machining were perfectly clean (free from grease or other soil), all common liquids would spontaneously and completely wet the surfaces. (The only exception would be observed when a metal became soiled by chemical reaction with the liquid itself.) It is thus evident that engineering metals are normally soiled because water, for example, will usually not spread on such metals. Metal surfaces freshly exposed in cutting are momentarily clean in the best sense; however, they quickly become soiled by reaction with--or adsorption of--materials from their gaseous or liquid environment.

Thus, their wettability changes, and because of such factors it is difficult to predict either the eventual degree or the ratio of their wetting from

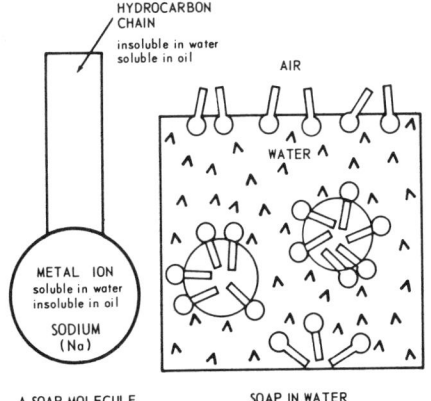

Figure 1-9. *Representation of surface active molecules in water. The adsorbed layer is "richer" in concentration than the body of solution-imparting wetting and lubricating effect. Soap is one example of thousands of available surface active products. (Courtesy, Master Chemical Corporation)*

known physical constants. Reductions in surface tension of water-based liquids will generally increase their overall wetting ability as seen in Figure 1-9; however, such recourses are not always successful.

One method avoided liquids and employed gases instead, as applied by Pahlitzsch (26) and others (21, 22, 23, 28) who used ambient or refrigerated streams of nitrogen, air, or carbon dioxide directed at high speed into the clearance space.

Another method, mist cooling, holds the promise of combining the advantages of gases (high velocity and good access) with those of liquids (utilization of heat of vaporization, plus inclusion of high molecular weight load-carrying agents). In this impingement approach, a high-velocity stream of aerosol droplets (diameter ≤ 1 micron) is driven into the cutting zone. Brosheer (29) reported success by this method. Shaw and Smith (30) found in tool life tests at 400 fpm (122 m/min), that a water mist gave less tool wear than a flood of water directed into the same area.

Another method involving impingement is the "highjet" technique of Pigott and Colwell (24), who applied a thin 0.015 inch (0.38 mm) stream of ordinary cutting fluid at high velocity into the cutting zone. Consistent decreases in wear rate over standard methods using the same fluid were observed. Practical problems developed with this approach however; the recirculation of machining-generated particulate matter resulted in plugging these orifices and necessitated extremely effective filtration to achieve practical success.

The primary access problems faced by fluids in the sense of wetting and penetration have been examined largely on the basis of simple metal surfaces. Obviously metal surfaces exposed during metal cutting are complex structures, and clearances at the tool tip and at the tool/chip interface are small as illustrated in Figure 1-10. The possible importance of a secondary phase of penetration by cutting fluids through these tiny openings must not be overlooked. Merchant (10), as a result of photomicrographic

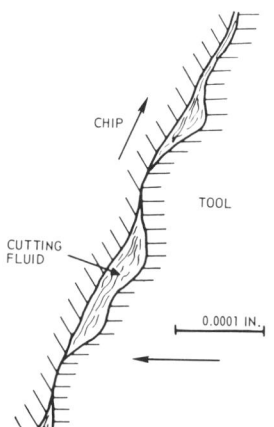

Figure 1-10. *Imaginary, highly magnified cross-section of small portion of chip/tool interface (31). (Courtesy, Cincinnati Milacron)*

evidence, talked in terms of capillaries of 0.015 inch (0.38 mm) diameter. He cited calculations to show that liquids with the viscosity and surface tension of water would be capable of penetrating such microcracks by capillary action.

In considering how chemical agents may achieve deep access into the shearstress zone, or how they may travel upward to the region of the tool/chip interface, requires acceptance of the capillary type of action. Otherwise, direct permeation of the metal by intergranular diffusion seems necessary to explain the facts.

Reports of precutting penetration of the workpiece by a load-carrying agent were cited by Low (32), although he was unable to verify them. The results, based on chip curl changes, indicated that the application of carbon tetrachloride to a workpiece before machining affected subsequent cutting to a depth of 0.02 inch (0.5 mm) below the original surface, regardless of the number of cuts taken. Cassin and Boothroyd (5) confirmed these effects. In the machining of copper, where carbon tetrachloride was applied only to the smooth top surface of the workpiece, a sudden increase in friction angle, ß, was seen as increasing depths of cut reached 0.015 inch (0.38 mm).

Usui, Gujral, and Shaw (33) obtained similar evidence of the ability of carbon tetrachloride to penetrate below the surface of an ordinary hot-rolled steel workpiece. The fluid, previously applied and allowed to dry, effected a reduction of cutting forces at slow speeds. Baking the workpiece instead of air-drying eliminated the effect. The effect also disappeared if the workpiece had been previously ground smooth, or if the cutting speed was quite high. This suggests that the initial penetration was through microscopic cracks in the metal surface which were removed by grinding. It also suggests that carbon tetrachloride effected penetration as the molecule--not as an iron reaction product like the chloride salt.

The results of these penetration studies imply that penetration is an important aspect of cutting fluid action, that penetration limitations at the molecular level are indeed significant, and that simple molecules are able to penetrate metal structures with signal effects on the cutting process.

THE COOLING MECHANISM

It was suggested earlier that high-speed cutting would produce the optimum result except for a few critical faults, notably drastic tool attrition due to heating effects. The obvious solution would be to cool the tool. Some experimental success has been obtained by this approach through the use of hollow tools with internal circulation of a cooling fluid (34). In the practical sense, however, this cannot be done, especially with brittle carbide and ceramic tools, because of structural and thermal conductivity factors. Therefore, from a practical standpoint, not just the tool but also the chip and workpiece must be cooled.

In the general cooling of the chip and tool in this manner, temperature reductions are registered in the cutting zone, with the effect of reducing plasticity and possibly increasing cutting forces (35). In addition, wear on the tool flank is usually not lessened. In certain cases, changes in chip curl cause increased flank wear (36) and there is no reason to expect consistent improvements in surface finish (7).

Operation of the cooling mechanism cannot be understood with certainty without carefully noting temperature changes effected by fluids at different locations near the cutting site. Calculations of local temperatures, particularly in the shear plane and at the tool/chip interface have been made by Rapier (37), Loewen and Shaw (38) and Chao and Trigger (39). None are fully satisfactory or highly reproducible. Direct methods, such as the tool/workpiece thermocouple method of Shore (40) and the thermosensitive color-technique of Schallbrock, et al., (41) are more realistic, but suffer similarly from variability of results and evidently from systematic error.

Few really convincing proofs of the uncomplicated ability of coolants to reduce cutting zone temperature are to be found in the literature, especially for high-speed cutting.

Cutting Temperature and Wear Effects

The primary reason for cooling is to retard high rates of face and flank wear by curbing the sharp temperature rise which accompanies short ranges of cutting speed increase. Therefore, either longer tool life for a given speed or higher speeds for a given tool life can be obtained. This is primarily due to a change in the constant as expressed in the Taylor equation (20):

$$Vt^n = \text{Constant} \qquad (1\text{-}2)$$

Where V = Cutting speed (fpm)
       t = Tool life (min)
      n = Exponent varying with tool and work

Other formulas involve temperature directly, like that of Cunningham and Phillips (42) for the machining of mild steel with different cutting fluids:

$$\emptyset \times 10^{-3} = t^{0.04} \qquad (1\text{-}3)$$

Where $\emptyset$ = °F at the tool/chip interface.

A study was carried out by Iyengar, *et al.*, (34) which compared the effect of an oil in water emulsion and a sulfurized oil with the action of an internally cooled cutting tool, and also with dry cutting (all tools were high-speed steel). Cutting temperature was judged by temper color of the chips and chip curl diameters. Cutting forces were also measured at two cutting speeds, 67 and 131 fpm (20 and 40 m/min). Low temperatures were correlated with smaller cutting forces and small chip curl diameters; higher temperatures were associated with higher cutting forces, and larger chip curl diameters. The results showed that cutting oils as well as aqueous coolants lowered cutting temperatures, the former indirectly by reducing heat generation, the latter by heat removal. The internally cooled tool produced similar results, but clearly through a thermal mechanism. It was concluded that the cooling effects of cutting fluids are of primary importance.

Dorinson (43, 44), has expressed doubt that any effective coolant action can be registered at the chip/tool interface due to the access problem. He contended that coolants lower shear zone temperature by removing heat along the back side of the chip as a gross effect of the flood of fluid over chip and workpiece. The resulting reduction in cutting temperature would influence tool life in accordance with an equation such as 1-3.

Shaw, Pigott, and Richardson (45) however, demonstrated the ability of aqueous fluids to perform as coolants in cutting SAE 1015 steel with high-speed tools at various depths of cut and speeds up to 400 fpm (122 m/min). The thermocouple cutting temperatures observed with two oil-in-water emulsions, a water-surfactant solution, and water itself were all about equal, and decidedly lower than in dry cutting. In this same study, water, benzene, and carbon tetrachloride were compared with dry cutting. At surface speeds of 100 to 425 fpm (30 to 129 m/min), water registered a temperature lowering of 200°F (111°C), while both organic fluids gave temperature-speed profiles very near to that for dry cutting.

In a study which cited average thermal properties vs. cutting temperatures, Eugéne (46) performed experimental machining involving two compounded oils, two aqueous fluids, and no fluid. The conditions and results of cutting at 50 fpm (15.25 mpm) are cited in Table 1-2. Surprisingly, the results show an almost linear relationship between cooling effect and specific heat in spite of the fluids' unquestioned differences in lubricating ability.

Shaw, Cook, and Smith (36), in experimental machining at 340 to 400 fpm (103.7 to 122 mpm) of AISI 1000 and 4340 steels with carbide tools, used thermocouples buried in the workpiece or attached to a zone that would become the chip. Workpiece thermocouples did register differences purported to reflect temperatures in the shear zone. With sharp tools, these temperatures were slightly *higher* with water than in dry cutting while, with tools having an appreciable wear land, they were slightly lower. The effect was judged to stem from chip curl resulting from rapid cooling which increased the temperature of both the lower part of the rake face (cutting

Table 1-2. Effect of Various Cutting Fluids in Lowering the Tool Workpiece Temperatures Measured When Machining Steels[1]

Cutting Fluid		Temperature Decrease, Percent[2]	
Composition[3]	Specific Heat BTU/(lb.)(°F.)	Steel A (tempered)	Steel B (annealed)
Compounded oil, low viscosity	0.489	3.9	4.7
Compounded oil, high viscosity	0.556	6	6
Aqueous solution of wetting agent	0.872	14.8	8.4
Aqueous "soda product" solution 4 per cent	0.923	–	13
Water	1.00	19	15

[1] Using high-speed tools at 50 sfpm., depth 0.19 in., feed 0.016 in., rake and clearance angles 20 degrees and 4 degrees, respectively, and cutting speed at 50 sfpm.
[2] Temperatures measured during dry cutting: Steel A, 1480°F.; Steel B, 1340°F., taken as 100 per cent.
[3] Commercial cutting fluids.

tip) and also the area near the wear land. It was concluded that aqueous fluids can be *too active* as coolants, with tighter curling of the chip giving undesirable wear on the rake face also. To overcome this undesirable effect under practical cutting conditions, feeds should be increased (36), as illustrated in Figures 1-11 and 1-12.

Hall (47) also found that coolants can give negative as well as positive effects during metal cutting, particularly with regard to tool wear and surface finish results. In a series of experiments while machining steel, he measured: (l) cutting ratio, (2) chip area compression, and (3) chip hardness as a function of temperatures 100° to 1200°F (37.8° to 648.9°C) by changing cutting speed. The cutting ratio (r_c) fell rapidly from a maximum at 300°F (148.9°C) to a minimum at 650°F (343.4°C), after which it rose again. Similarly, chip hardness was maximum and area compression minimum at about 700°F (371°C) above which these values fell and rose respectively at very high rates.

Thus, it was concluded that, if the temperature were either lowered below the 600°F (315.6°C) region, or allowed to rise significantly above it, the effect on both cutting ratio and chip area compression would be undesirable and chip brittleness would be reduced. The optimum cutting temperature in this sense, however, corresponded to a maximum in the surface roughness curve.

In comparisons among aqueous fluids, a cutting oil, and dry cutting, the oil was apparently able to maintain such optimum temperature conditions, plus giving better finish and tool life. At extreme speeds, above 300 fpm (91.5 mpm), however, the aqueous coolant prevented the catastrophic drop in tool life seen with both oil and dry cutting.

It would seem that, while the use of lubricating cutting oils can be relied upon to give either positive effects, or none, judgment and care must be

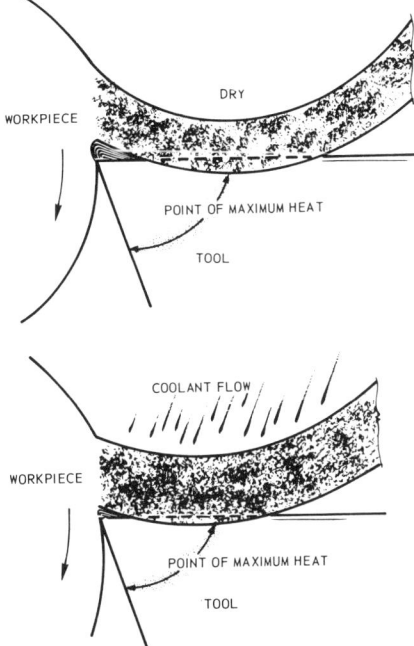

Figure 1-11. *Showing the area of greatest heat concentration as half-way between tool tip and the point where the face of the chip leaves the tool face. (Courtesy, Master Chemical Corporation)*

Figure 1-12. *The same speeds and feeds compared with Figure 1-11. Chip curl is tighter, concentrating heat nearer the cutting edge and reducing tool life. To correct condition, increase feed compared with feed used cutting dry. (Courtesy, Master Chemical Corporation)*

used in the application of aqueous fluids (including chemical solutions and emulsions) as coolants, to avoid negative results. It is clear, however, that the coolants are uniquely active at high cutting speeds where lubricant action is ineffective and where chip-tool temperature reduction is very important to tool life.

Cutting Fluid Temperature Effects

The temperature and tool-life effects of fluids with different innate cooling abilities have been reviewed. The cutting mechanism can be further explained by examining the effect of supplying refrigerated fluids. If colder fluids show any particular merit, their action as coolants is confirmed.

Boston, Gilbert, and McKee (48) demonstrated the involvement of this factor in studies on a 5% oil/water emulsion as against the oil itself (a sulfurized mineral oil). The fluids were supplied at 40° to 150°F (4.5°-65.5°C) at a constant rate during the machining of annealed SAE 3140 steel with high-speed tools; cutting speeds were 100 to 200 fpm (30.5 to 61.0 m/min). Both fluids gave better tool life than with dry cutting at 100 fpm (300 mpm) but were more effective when cold. At 200 fpm (600 mpm), the best tool life was registered by the coldest fluid 40°F (4.4°C), but the relative effectiveness of the emulsion was lower than that of the oil. The negative effect of increased viscosity at lower temperatures was evident, particularly for the oil.

An example not complicated by viscosity limitations was given by Pahlitzsch (26) who compared refrigerated with ambient compressed air as a coolant in cutting at 100 fpm (30.5 m/min). At the lower supply temperatures, -40 to -76°F (4 to -60°C), a fourfold improvement in tool life was achieved compared with that at 18° F (-7.8°C). Similar results where obtained with CO_2.

In further work (49), the same author compared refrigerated 34 to 39°F (1 to 4°C) and ambient 108°F (42°C) liquid coolants for the deep hole drilling of steel tubular members using high-speed bits. The cutting fluid was a commercial 8% oil/water emulsion. Tool wear rate showed a 2.5 times decrease at the lower temperature; both power requirements and workpiece temperatures were considerably reduced. Similar results (50) were obtained in the milling of gear wheels at 260 fpm (79 m/min), which resulted in a 2.9% net decrease in machining cost.

The various applications of nitrogen as a gaseous cutting fluid, and particularly those of CO_2 gas and CO_2 mist, are examples of the value of chemically inert materials whose action is evidently exerted by the cooling mechanism.

Bickel (51) conducted cutting experiments which revealed the relative importance of cooling action at very high cutting speeds. He compared the effect of a CO_2 mist (delivered at -110°F [-43°C]) with that of carbon tetrachloride and a commercial cutting oil vs. dry turning at cutting speeds from 100 to 600 fpm (30.5 to 183 m/min). His criterion of cutting fluid activity was increased in shear angle (see Figure 1-1).

His results, shown in Figure 1-13, indicate the superiority of carbon tetrachloride over the inert coolant at speeds below 250 fpm (76 m/min) where, however, both coolants and lubricants are superior to dry cutting.

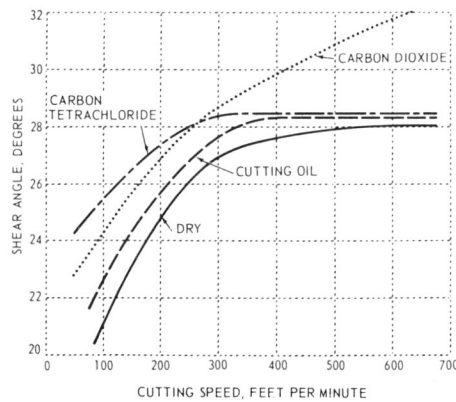

Figure 1-13. *Relationship between shear angle and cutting speeds for different cutting fluids.*

Above that speed, the relative superiority of the coolant becomes more pronounced. At the highest speeds, the chemically active compounds register effects little better than that in dry cutting; in this range, obviously, only the cooling mechanism is operative.

A recent very detailed comparison (52) of the machining of super-strength alloys with refrigerated gaseous and liquid coolants vs. dry refrigerated (-80°F [62.2°C]) ambient machining is probably the best available demonstration of the cooling mechanism in practical use. The alloys included:

A-286 (austenitic stainless steel)
L-605 (cobalt base high temperature alloy)
R-235 (nickel base precipitation hardening alloy)
H 11 (high-strength hot-work steel)
Mo-0.5% Ti (high-temperature high-strength alloy)

The fluids were CO_2 mist (-74°F [-58.9°C]) and a mixture of 25% trichloroethylene and 75% petroleum naphtha at -50°F (-45.5°C). Turning and milling, using high-speed and carbide tools, were performed on regular commercial equipment.

Tool life, cutting forces, and economic factors were observed using a range of cutting speeds from 70 to 650 fpm (21 to 198 m/min). Cutting temperatures up to 2000°F (1093°C), normally observed in dry cutting, were reduced significantly by the CO_2 mist and the refrigerated liquid coolant to give as much as a threefold improvement in tool life. Cutting forces were measurably reduced in many cases, particularly with the liquid coolant.

The results with L-605 cobalt base steel, shown graphically in Figure 1-14, are typical. In turning experiments using carbide tools, both CO_2 mist and the liquid coolant gave very large increases in tool life in the range between 130 to 190 fpm (39 to 57 m/min); tool life was disastrously low with dry cutting (as low as 155 fpm [46.5 m/min]). Both tangential and normal cutting forces were understandably higher with the coolants, although with some metals, (notably R-235 nickel-base steel and the molyti alloy) greater tool life was achieved without the expense of greater work.

For comparison, workpieces were precooled to -110°F (-43.4°C) in a cold chest and removed just before machining. This method generally gave improved tool life also, but the effects were much less pronounced. The similarity between the results with the inert CO_2 mist and the fluid containing trichloroethylene indicates the action observed in both cases was exerted by the cooling mechanism, rather than chemical action as in extreme pressure lubrication.

The foregoing discussion of experiments involving the application of cold cutting fluids, has served to isolate the results of effective lowering of temperatures in the cutting zone. The important effects of such action, notably tool life increases, were clearly examples of cutting fluid perfor-

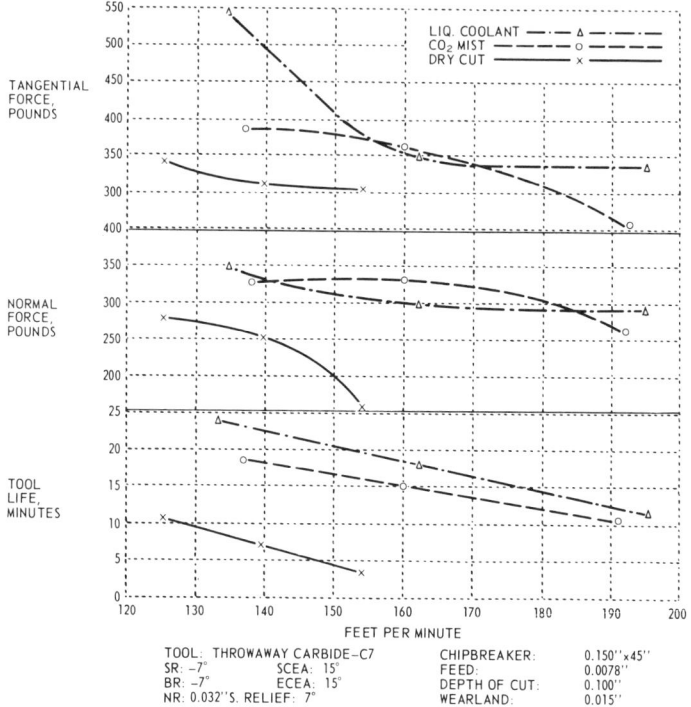

Figure 1-14. *Comparison of tool life results and cutting force measurements in turning L-605 steel: dry, with CO_2 mist, and with refrigerated liquid coolant (52).*

mance by the cooling mechanism. It seems to be true that while heat generation in cutting can be reduced beneficially by lubrication, or its equivalent, it cannot be wholly eliminated. Thus, the considered choice of fluids with superior cooling abilities is likely to remain important in machining technology.

THE LUBRICATION MECHANISM

Lubricants reduce friction between surfaces which are in relative motion. At any given rate, the force holding the parts together controls the intimacy of contact and thus has a direct influence on frictional force. When normal forces are quite low, the viscosity of an intervening layer of fluid can be the major factor in regulating friction (53).

In metal cutting, where normal contact stresses at the chip/tool interface may exceed 50,000 psi (344 kPa) (54), 10^{-11} inches has been estimated (55) as the maximum thickness of fluid films with the low viscosities found at cutting temperatures such as 900°F (482.2°C). Since the average surface asperity height is 10^{-3} to 10^{-4} inches, it seems certain that fluid-film lubrication cannot exist under these conditions.

Supposedly "smooth" metal surfaces are composed of hills and valleys. Static contact of such surfaces consists of a degree of mating of these irregularities when a load (normal force, N) is applied. Greater intimacy of contact is achieved by deformation of the contacting asperities, and the real area of contact increases until the compressive yield strength of the metal in actual contact is equal to the load. If a tangential force, T, is applied, sliding occurs at a value of T given by the equation:

$$\mu = T/N \qquad (1\text{-}4)$$

Where μ is the coefficient of static friction.

This is Amontons' Law, which governs sliding friction--a situation where the real area of contact is: (l) small compared to apparent area, and (2) proportional to the normal force.

Under conditions of high temperature and pressures in the contact area, however, a degree of asperity adhesion due to welding (local melting) of asperities occurs, and force is required to break such adhesions. The apparent coefficient of friction under these conditions rises, but the requirements for Amontons' Law to apply still prevail if only a small proportion of the apparent surface is involved.

This is the physical realm where boundary and extreme-pressure lubrication are important. Chemical lubricants are useful under these conditions. The gouging, ploughing, and transfer of metal from one surface to another is opposed by physically absorbed or chemically interposed films, respectively. Such films must be readily sheared at temperatures and pressures lower than those at which the shearing of asperities would normally occur to obtain beneficial effects. Such components in the past have been referred to as conferring "oiliness," involving modification of the metal surface. Boundary lubricant ingredients, like fatty acids, give films at ambient temperatures, largely by adsorption.

Secondary action by these materials, and primary action by extreme-pressure (EP) lubricants, can involve chemical reaction to form surface layers of metal salts (56). These reactions may, especially with chlorine and sulfur compounds, occur at those temperatures at which their lubrication effects are required; thus, differences in temperature thresholds and subsequent film-forming reaction rates are important. The effectiveness of these secondary (chemical) films is understood to be limited by their melting points: fatty soaps up to 212°F (100°C), iron chloride to 1112°F (600°C), and iron sulfide to 1832°F (1000°C) (57). Above their respective melting points, in accordance with the above interpretation of their function as solid lubricants of low shear strength, only ineffectual fluid-film action could remain, and they would cease to function (58).

This interpretation of EP lubricant action is consistent with the main stream of theoretical viewpoint, as expressed by Bowden and Tabor (56). Details of the actual operation of EP lubricants varies in the opinions of Groszek (59), Rabinowicz (60), and Kohn (61). It is theorized that much of the favorable effects of EP lubricants in metal cutting may stem from the changes they induce in the deformability of the metal itself, under compressive stresses in metal cutting.

It is well established that boundary and EP ingredients in cutting fluids not only reduce cutting forces but can also improve surface finish and provide longer tool life (62). Most petroleum chemists (57, 63, 64, 65, 66, 67) find it natural to accept the familiar precepts of boundary and EP action as a basis for their consideration of cutting oils.

The relationship seems a very close one. For example, Brookman and Ham (68) have studied a series of fatty oils, sulfurized compounds, and chlorine derivatives in both sliding-friction tests and simple metal cutting experiments. The results gave a reasonably close correlation between the varying efficiencies in these two kinds of action. In recognition of this fact, the 4-Ball Tester, designed for EP lubricant studies, serves as a laboratory test for screening possible cutting-fluid additives (7).

Colloidally dispersed solid lubricants such as molybdenum disulfide and graphite are known to be useful as antiweld agents in EP lubricants, with effects resembling that of iron sulfide (a reaction product of dissolved active sulfur compounds) in the same application. These low shear-strength solids also give superior lubrication effects in metal cutting (69, 70, 71).

THE SHEAR-STRENGTH-REDUCTION MECHANISM

During the past 30 years, Russian investigators have observed the effect of adsorbed chemical agents upon the characteristics of surfaces. Action of this type, as in the reduction of effective hardness in metals, has been termed the "Rebinder Effect." It was invoked (72) as early as 1936 as a basis for the "oiliness" of boundary lubricants. Rebinder (73) observed, in tension studies of tin specimens, that a boundary-lubricant film of oleic acid both lowered the flowstress curve and increased the observed number of active slip planes. He concluded that this resulted from penetration of the surface active acid into microcracks and that it lowered the strength of the specimen. He termed this "hardness reduction." In a slightly different viewpoint of the overall weakening effect, Pleteneva, Shreiner, and Rebinder (74) reported an acceleration of strain-hardening when metal was periodically stressed in the presence of adsorbed surfactants. The hardening of individual elements was suggested (75, 76) as weakening the metal matrix generally so that shear could occur at reduced levels of stress.

Rebinder's studies were directly concerned with the metal cutting process, and he performed many cutting experiments (77, 78, 79, 80). This work included the observation of microcrack formation in the metal near the tool tip, the smallest of which can quickly heal. The adsorbed film was believed to prevent the closing of these microcracks as the load changed during cutting. The increased number of zones of stress concentration yielded a closer spacing of active shear planes, with the result of smoother cutting at lower energy expense.

Epifanov distinguished between "adsorptional plasticizing" of the metal surface (hardness reduction) and the more important depth effect, the intensification of strain-hardening of individual metal elements which produced a tendency toward structural disintegration in response to compressive stress. In metal cutting, he observed that cutting fluid activity can be experimentally related to the length of individual slip elements in the chip. This observation has been affirmed by Shaw (81). A higher decrease in slip element length does correlate with decreased cutting forces and improved surface finish. Epifanov, in perhaps an unfortunate choice of terms and analogies, however, suggested that the penetration of foreign atoms (from cutting fluid decomposition) produced an embrittlement effect in a manner similar to hydrogen embrittlement. He concluded that cutting is thus facilitated by a resulting decrease in plasticity.

Thus, a debate prevails in the interpretation of the role of carbon tetrachloride and similar "chemically active" cutting fluids. The details have not yet been filled in, but it seems possible that most of the effects formerly ascribed to EP lubricant action on the part of chlorine, sulfur, and other boundary and EP types of fluids, will emerge as primarily due to shear-stress reduction of the workpiece metal during machining.

Reappraisal of cutting forces is consistent with the shear strength reduction mechanism. When active chemical fluids enter the cutting zone, they produce and stabilize a uniform network of surface imperfections. This results in more active slip planes, and an associated secondary flow within the chip. This secondary flow can only be effective, however, if strain-hardening of the chip surface can promote chip curl, and it is materially aided by a primary increase in tool face friction resulting from presence of reaction products like ferric chloride. These factors can be responsible for reductions in apparent tool face friction coefficient, increases in shear angle, and reduced cutting forces.

INTERRELATIONSHIP OF MECHANISMS

The cooling mechanism has been shown to operate independently of what has usually been considered the lubricating mode of chemical action.

Both depend on physical access to the cutting zone, although probably to different degrees. In addition, the cooling and chemical actions of a given fluid may have simultaneous effects in opposite directions under some cutting conditions. This was made particularly evident by the work of Hall (47) and Shaw (36), and it seems apparent that cooling the chip is as undesirable in slow-speed cutting as it is essential at extreme speeds. The major indicators of good balance between the two is the achievement of good surface finish at low speeds and acceptable tool life at high speeds.

The problem of reconciling the lubricant and shear strength lowering mechanisms is more complex because they are, perhaps, not only interrelated but to an extent, identical. Kohn (61) has presented a revised view of EP lubrication which, in essence, suggests that such lubrication is effected through the promotion of shear by stabilizing microcracks in junctions established under the high normal stresses that occur there, just as in metal cutting.

Shaw (81) has emphasized that an essential secondary aspect of the chemical action of fluids via the new mechanism is antilubrication at the tool/chip interface, and a consequent strain-hardening of metal in the chip surface to cause curl. The primary aspect, however, is the microcrack stabilization, which leads to reduction of shear resistance throughout the zone of deformation. Shaw relates the resulting creation of a larger number of uniformly spaced slip planes to increased plasticity, not to embrittlement, as suggested by the Russian researchers.

The embrittlement concept apparently originated in Rebinder's study of changes in relaxation of single crystals in which a coating of surface active agents appeared to cause formation of microcracks. Creation of such defects in the crystal lattice corresponds to transgranular fracture in an ordinary polycrystalline metal matrix, and this would certainly correspond to the effects of embrittlement.

If, however, it is assumed that chemical action is normally registered by facilitating intergranular slip, the change effected is better considered as an increase in plasticity. In the cutting example, $FeCl_3$, formed by chemical reaction after CCl_4 penetration, may be considered as altering the material of the grain boundary to facilitate slip.

In a sense, then, it may be true that the effects of active atoms like chlorine and sulfur as registered in metallic strength reduction may be very closely related to their previously supposed function as EP lubricants. The lubrication mechanism may therefore differ only in detail and not in substance from the revised viewpoint of many American researchers. This view, according to Shaw (81), combines the chemical reaction aspect of the former EP lubrication mechanism and accepts the homogeneous strain aspect of the Russian embrittlement mechanism.

REFERENCES

1. W. H. Northcott, "A Treatise on Lathes and Turning," (London: Longmans Green and Company, 1868).
2. A. Mallock, "The Action of Cutting Tools," Proc. Royal Soc., 33 (1881), p. 127.
3. M. E. Merchant, "Orthogonal Cutting and a Type 2 Chip," J. Appl. Phys., 16 (1945), pp. 267-75.
4. M. E. Merchant, "Plasticity Conditions in Orthogonal Cutting," J. Appl. Phys., (1945), pp. 318-24.
5. C. Cassin and G. Boothroyd, "Lubricating Action of Cutting Fluids," J. Mech. Eng. Science, 7, No. 1, (March, 1965), pp. 67-81.
6. E. G. Thomsen, J. T. Lapsley, and R. C. Grassi, "Deformation Work Absorbed by the Workpiece During Metal Cutting," Trans. ASME, 71 (1953), p. 591.
7. I. M. Feng, A. Gujral, and M. C. Shaw, "Cutting Fluid Performance," Lub. Eng., 17 (July, 1961), pp. 324-9.
8. W. B. Heginbotham and S. L. Gogia, "Metal Cutting and the Built-Up Nose," Proc., I. Mech. E., 175 (1961), p. 892.
9. Hans Ernst and M. E. Merchant, "Chip Formation, Friction and High Quality Machined Surfaces," Trans. ASM, 29 (1941), pp. 299-378.
10. M. E. Merchant, "The Physical Chemistry of Cutting Fluid Action," Am. Chem. Soc., Div., Petrol Chem., Preprint 3, No. 4A, (1958), pp. 179-89.
11. K. Okushima and K. Hitomi, "On The Cutting Mechanism for Soft Metals," Memoirs of the Faculty of Engineering. Kyoto University, 18 (1957), pp. 135-66.
12. K. J. Trigger and B. J. Chao, "An Analytical Evaluation of Metal-Cutting Temperatures," Trans. ASME, 73 (January, 1951), pp. 57-68.
13. K. J. Trigger, "Progress Report No. 2 on Tool-Chip Interface Temperatures," Trans. ASME, 71 (February, 1949), pp. 163-74.
14. J. C. Herbert and L. Fersing, "Research in Metal Turning," Western Machinery and Steel World (November, 1955).
15. R. C. Fisher, "Grinding Dry With Water," Grinding and Finishing, 11, No. 3 (March, 1965), pp. 32-4.
16. L. P. Tarasov, "Grinding Fluids," The Tool and Manufacturing Engineer 46, No. 7 (June 1961), pp. 67-73.
17. Ibid., 47, No. 1 (July, 1961), pp. 60-67.
18. Ibid., 47, No. 2 (August, 1961), pp. 57-63.
19. D. A. Stewart and H. R. Soderstrom, "Residual Stress vs. Cutting Fluid Selection When Grinding High Temperature Alloys," Lub. Eng., 17, No. 6 (June, 1961), pp. 286-90.
20. F W. Taylor, "On the Art of Cutting Metals," Trans., ASME, 28 (1907), pp. 31-350.
21. E. J. Tangerman, "Carbon Dioxide Cools Tools," American Machinist, 96 (March 3, 1952), p. 129.
22. A. I. Icaev and N. I. Tashlitskiy, "Use of Carbon Dioxide for Increasing Tool Life," Vestnik Mashionstroyeniya, 35 (November, 1955), pp. 42-8.

23. N. P. Sobolev and A. M. Z. Valitov, "Use of Carbon Dioxide for Turning Steels of Low Machinability," *Stanki Instrument*, 27 (March, 1956), pp. 18-21.
24. R. J. S. Pigott and A. T. Colwell, "Hi-Jet System for Increasing Tool Life," *SAE Quarterly Trans.*, 6 (July, 1952) pp. 547-66. Abstracted in *SAE Journal*, 60 (April, 1952), pp. 45-8.
25. M. C. Shaw, "Chemico-Physical Action of a Cutting Fluid," (D.Sc. thesis, University of Cincinnati, 1942).
26. G. Pahlitzsch, "Gases Are Good Cutting Coolants," *American Machinist*, 97 (February 16, 1953), pp. 196-97.
27. G. M. Hain, "Measuring the Cooling Properties of Cutting Fluids," *Trans. ASME*, 74 (1952), p. 1077.
28. M. C. Shaw, *Metal Cutting Principles*, (3rd edition, Cambridge, Massachusetts: MIT Press, 1960).
29. B. C. Brosheer, "Mist Cooling," *American Machinist*, 97 (September 14, 1953), pp. 137-52.
30. M. C. Shaw and P. A. Smith, "Evaluation of Mist Application of Cutting Fluids," *ASTE Research Fund Report No. 4*, (1956).
31. Hans Ernst, "Fundamental Aspects of Metal Cutting and Cutting Fluid Action," *Annals of New York Acad. of Sciences, Series II*, 53, No. 4 (June, 1951), pp. 805-23.
32. A. H. Low, "The Effect of Carbon Tetrachloride in Metal Cutting," *N.E.L. Report*, 18 (1962).
33. E. Usui, A. Gujral, and M. C. Shaw, "Experimental Study of Action of CCl_4, in Cutting and Other Processes Involving Plastic Flow," *Int. J. Machine Tool Design and Research*, 1, No. 3 (November, 1961), pp. 187-97.
34. H. S. Iyengar et al., *ASME Paper 64-Prod. 2 (April 20-24, 1964)*.
35. F. W. Boulger, "What Is Known Today about Metal Cutting," *ASTE Research Report*, No. 14, Paper No. 44 (May 1, 1958).
36. M. C. Shaw, N. H. Cook, and P. A. Smith, "Cooling Characteristics of Cutting Fluids," *ASTE Research Report 19*, (September 15, 1958).
37. A. C. Rapier, "A Theoretical Investigation of the Temperature Distribution in the Metal Cutting Process," *British J. of Appl. Phys.*, 5 (November, 1954), pp. 400-05.
38. E. G. Loewen and M. C. Shaw, "On the Analysis of Cutting-Tool Temperatures," *Trans. ASME*, 76 (February, 1954), pp. 217-31.
39. B. T. Chao and K. J. Trigger, "The Significance of the Thermal Number in Metal Machining," *Trans. ASME*, 75 (January, 1953), pp. 109-20.
40. H. Shore, "Tool and Chip Temperatures in Machine Shop Practice," (M.S. thesis, MIT, 1924).
41. H. Schallbrock, H. Schaumann, and R. Wallichs, "Testing for Machinability by Measuring Cutting Temperature and Tool Wear," in "Vortage der Hauptversammlung der Deutschen Gesellschaft fur Metallkunde," *VDI Verlag*, (Berlin: 1938), pp. 34-8.
42. D. M. Cunningham and R. E. Phillips, "The Tool-Life, Tool-Temperature Relationship as Affected by Cutting Fluids," (thesis, University of Cincinnati, 1951).
43. A. Dorinson, "A Theory of Cutting-Tool Wear and Cutting-Oil Action," *Trans. ASLE*, 1 (1958), pp. 131-38.

44. _____, "Crater Wear and Failure of High Speed Steel Tools in Dry and Lubricated Cutting," *Trans. ASLE, 6*. No. 4 (October, 1963), pp. 270-5.

45. M. C. Shaw, J. D. Pigott, and L. P. Richardson, "The Effect of the Cutting Fluid Upon Chip-Tool Interface Temperature," *Trans. ASME, 73* (January, 1951), pp. 45-56.

46. F. Eugéne, "New Method for Evaluating Coolant Efficiency," *Microtecnic, 9*, No. 2 (1955), pp. 70-80.

47. H. L. Hall, "Ein Beitrag zur Abgrenzung der Wirkungen von Kuehlung und Schmierung eines Kuehischmiermiteis auf die Zerspanung von Stahl," *Werkstattstechnik, 49*, No. 5 (May, 1959), pp. 247-54.

48. O. W. Boston, W. W. Gilbert, and R. E. McKee, "Influence of Applying Cutting Fluids at Different Temperatures When Turning Steels," *Trans. ASME, 67* (April, 1945), pp. 217-24.

49. C. Pahlitzsch, "Influence of Low Temperature Cooling on Tool Life in Deep Drilling with Twist Drills" *Microtecnic 8*, No. 6 (1954), pp. 327-31.

50. _____, "Low Temperature Cooling as a Means of Increasing Cutting Tool Life," *Microtecnic, 9*, No. 2 (1955), pp. 65-9.

51. E. Bickel, "Contribution to the Theory of Frictional Wear of Turning Tools," *Microtecnic, 9*, No. 2 (1955). pp. 53-8.

52. F. A. Monahan et al., *American Machinist/Metalworking Manufacturing, 104*, No. 10 (May 16, 1960), pp. 109-24.

53. O. Reynolds, "On the Theory of Lubrication and Its Application to Mr. Beauchamp Tower's Experiments," *Phil Trans. Roy. Soc., 177*, pp. 157-234.

54. The Texas Company, "Some Recent Concepts of Machinability," *Lubrication, 38* (July, 1952), pp. 81-96.

55. E. E. Bisshopp. E. F. Lype, and S. Raynor, "The Role of Cutting Fluid as a Lubricant," *Lub. Eng., 6* (April, 1950), pp. 70-3.

56. F. P. Bowden and D. Tabor, *The Friction and Lubrication of Solids*, Chapter I (Oxford University Press, 1954).

57. A. L. H. Perry, "Lubrication in Metal Working," *Metalurgia, 48* (July, 1953), pp. 3-10.

58. A. W. J. Chisholm and W. S. McDougall, "Progress Report on an Experimental Investigation into the Effect of Strain Hardening in the Mechanics of Orthogonal Cutting," *M.E.R.L. Plasticity Report*, No. 61 (1952).

59. A. J. Groszek, "Heat of Preferential Adsorption of Surfactants on Porous Solids and Its Relation to Wear of Sliding Steel Surfaces," *Trans. ASLE, 5* (1962), p. 105.

60. E. Rabinowicz, "Influence of Surface Energy on Friction and Wear Phenomena," *J. Appl. Phys., 32* (1961), pp. 1440-444.

61. E. M. Kohn, "Role of Extreme-Pressure Lubricants in Boundary Lubrication and in Metal Cutting," *Nature, 197* (1966), p. 895.

62. C. A. Sluhan, "How Cutting and Grinding Fluids Effect Value Analysis-Manufacturing Cost Reduction," *ASTME Technical Paper SP63-126, Creative Mfg. Seminars*, (March, 1963).

63. S. J. Beaubien and A. G. Cattaneo, "A Study of the Role of the Cutting Fluid in Machining Operations," *Lub. Eng., 10* (April, 1954), pp. 74-79.

64. L. H. Sudholz, "Cutting Fluids: Fundamentals and Laboratory Evaluation," *Lub. Eng., 13*, No. 9 (September, 1957), pp. 509-15.

65. A. Dorinson, "Concentration Effects of Cutting Oil Additives in Performance Evaluation," *Lub. Eng.*, 12 (November-December, 1956), pp. 387-91.

66. H. A. Hartung, J. W. Johnson, and A. C. Smith, Jr., "Threading Test for Cutting Oil Evaluation," *Lub. Eng.*, 13, No. 10 (October, 1957), pp. 538-42.

67. J. M. Stokeley, "A New Performance Test for Cutting Fluids," *Lub. Eng.*, 9 (June, 1953), pp. 137-39.

68. J. G. Brookman and R. B. Ham, CSIR, (Australia) *Tribophysics Division Report* A-37 (1941).

69. A. Niedzwiedzki, "Lubrication of the Cutting of Metals," *Machinery* (London) 83 (October 9, 1953), pp. 717-21, 741.

70. *Ibid.*, 84 (February 12, 1954), pp. 337-43.

71. *Ibid.*, 85 (August 6, 1954), pp. 280-86.

72. B. Deryagin and Kusakov, "The Characteristics of Thin Films and Their Influence on the Interaction of Solid Surfaces," *Isvest, Akademii Nauk. CCCR*, 5 (1936), pp. 741-52.

73. P.A. Rebinder and V.I. Likhtman, "Effects of Surface-Acting Media on Strains and Ruptures in Solids," *Proc. 2nd Int. Cong. of Surf. Activity*, (London: Butterworths, 1947), pp. 563-80.

74. N.A. Pleteneva, L.A. Shreiner, and P.A. Rebinder, "Strain Hardening of Aluminum During Cutting in Inactive and in Surface Active Cutting Fluids," *Doklady Akad, Nauk., CCCR*, 62 (1948), pp. 653-55.

75. G.I. Epifanov, N. A. Pleteneva, and P. A. Rebinder, "Mechanism of Action of Active Agents in the Cutting of Metals," (Brutcher Translation), *Doklady Akad. Nauk., CCCR*, 97, No. 3424 (1954), pp. 277-79.

76. _____, "Influence of Surface Active Agents on Chip Elements," (Brutcher Translation), *Doklady Akad. Nauk., CCCR*, 104, No. 3719, pp. 68-71.

77. _____, P. A. Rebinder, and L. A. Shreiner, *Doklady Akad, Nauk., CCCR* 64 (1949), p. 879.

78. _____ and L. A. Shreiner, *Doklady Akad. Nauk., CCCR*, 80 (1951), p. 781.

79. _____, F. P. Soloshka, and P. A. Rebinder, *Doklady Akad, Nauk., CCCR*, 99 (1954), p. 801.

80. V. D. Kuznetsov, "Surface Energy of Solids," HM Stationary Office, (London: 1957).

81. M. C. Shaw, "On Action of Metal Cutting Fluids at Low Speeds," *Wear-Usure-Verschleiss*, 2, No. 3 (February, 1959), pp. 217-27.

CHAPTER 2
FLUID TYPES

F. W. Taylor (1) was one of the first to prove the practical value of using liquids to aid in metal cutting. In 1883, he demonstrated that a heavy stream of water, flooding the tool/chip/workpiece contact area during cutting, increased cutting speed by 30 to 40%. The early findings of Taylor and others led to the development and use of fatty oils for all types of metal cutting.

Further improvements followed quickly. Mineral oils were developed for machining brasses and other nonferrous alloys, and for light operations on steel. Mineral oil and lard oil blends proved effective for the more difficult cutting operations, and met machining requirements at that time.

However, the demand for increased production accompanying the spread of industrialization spurred efforts to increase cutting speeds and develop cutting tool materials of greater hardness and toughness. The development of such tool materials created a corresponding need for improved cutting fluids. Extensive research by fluid manufacturers and users alike, produced special chemical additives and other means to impart the combinations of qualities required by heavyduty machining.

Today, many blends and compounds provide cutting and grinding fluids for every modern machining requirement. Fluids may be a straight mineral oil, or a mixture of oils, and may contain one or a combination of additives such as sulfur, chlorine, phosphorus, or other chemicals. In addition, the emulsifiable or water miscible oils are widely used. Mixed with water, they form emulsions for use in machining and grinding where the primary need is for a moderate coolant rather than a lubricant. More recently, water miscible fluids using less oil (or no oils), and based on chemicals with or without surface active agents, have provided industry with products of greater heat conducting properties for still higher machining rates. Developments incorporating extreme pressure and oiliness additives give cutting fluids the necessary properties for application to a wide variety of machining operations.

With the choices available, no cutting or grinding operation need be hindered by a lack of proper fluid. This section contains information on the composition, function, advantages, and disadvantages of the fluids available to the metalworking industry.

CUTTING OILS

A plain cutting oil is an oil derived from petroleum, animal, marine, or vegetable origin, used straight or in combination.

Mineral Oils

Uncompounded mineral oils used in their natural state or as a component of cutting and grinding fluids can be divided into two general categories: *naphthenic* mineral oils and *paraffinic* mineral oils. Basically, the naphthenic mineral oils have saturated, ring-type structures; the paraffinic mineral oils are straight, or branched-chain saturated hydrocarbons.

The application of naphthenic and paraffinic mineral oils in a straight, uncompounded form is restricted to very light-duty applications on metals of high machinability, such as aluminum, magnesium, brass, and sulfurized, or leaded free-cutting steels. They are satisfactory only as hydrodynamic or fluid-film type lubricants. Mineral oils must be compounded with surface reactive additives if they are to be used as boundary or extreme pressure lubricants.

While straight mineral oils are still used as cutting oils, their primary function is as a blending medium, or a carrier of additives in cutting oils and water miscible fluids. When used as such, the naphthenic oils are preferable to the paraffinic because they aid in obtaining a more homogeneous compound. Compared with naphthenic, the paraffinic oils have superior resistance to oxidation and deteriorization at high temperatures. They are not, however, as easily miscible with additives.

Polar Additives. These are oils, fats, certain waxes, and synthetic materials added to mineral oils to increase their load-carrying capacity, or cutting capabilities. The fats, fatty oils, waxes, etc., may be of animal, vegetable, or marine origin. Polar additives find many uses in water miscible and oil-phase cutting fluids. In cutting oils, they may be dispersed as is, or reacted with sulfur-and/or chlorine-bearing chemicals, then dispersed in the carrier oil. In water miscibles, they are most often used in the form of soaps and fatty-amine or amide condensates.

Many synthetic polar additives are also used in cutting fluids as partial boundary lubricants. Some of the most common are esters, condensates or fatty oils and fatty acids, and poly or complex alcohols.

The function of any fat or oil polar additive is to wet and penetrate the chip/tool interface by reducing the interfacial tension between the carrier mineral oil and the metal. This is accomplished by polarity, or affinity of the polar additive for the metal substrate. The heat generated by the metal removal causes the absorbed film to react with the substrate metal, and the polar materials are converted to a low-shear organo-metallic film. This film provides lubrication by reducing the friction at the chip/tool interface. In essence, it acts as a partial boundary lubricant. Figure 2-1 shows the actual chemical structure of a saturated nonpolar hydrocarbon. Figure 2-2 shows an unsaturated polar organic molecule (oleic acid).

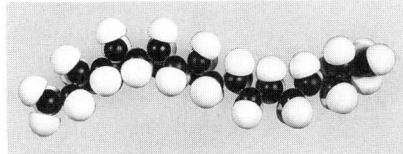

Figure 2-1. *Non-polar hydrocarbon. (Courtesy, Master Chemical Corporation)*

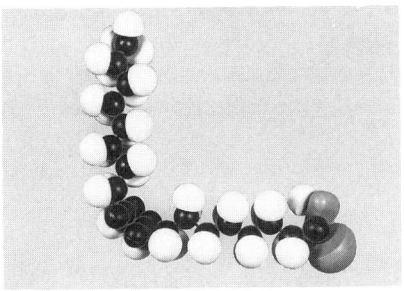

Figure 2-2. *Unsaturated polar organic molecule (olelic acid). (Courtesy, Master Chemical Corporation)*

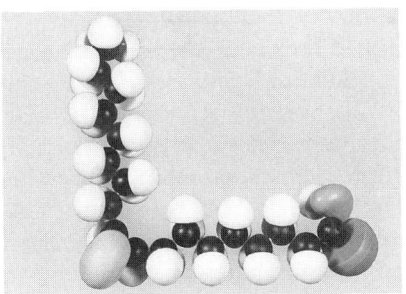

Figure 2-3. *Extremely pressure additive-sulfurized oleic acid. (Courtesy, Master Chemical Corporation)*

Unfortunately, the melting point of this film is rather low and its toughness is limited. Therefore, in very severe operations, it must be further supplemented by the use of extreme pressure additives such as sulfur and chlorine. Figure 2-3 shows an extreme pressure sulfurized polar organic molecule (oleic acid).

Because of their affinity for a metallic surface, the polar additives are also valuable as corrosion inhibitors. They lay down a tenacious "polar film" which offers a "barrier" of protection to the atmosphere. Although the metallic soap film does supply some boundary or extreme pressure lubrication, it is the least effective of the commonly used extreme pressure lubricants. This is due primarily to its relatively low melting point.

Animal. Animal fats and oils used as polar additives are derived from the fatty tissue of such animals as cattle, sheep, and swine. Due to their unsaturated chemical structure, the *oils* remain in liquid form at room temperature. The *fats*, however, are "saturated," and remain solid or semi-solid at room temperature. Fats and oils are good boundary lubricants. They can contribute to the reduction of frictional heat in cutting or grinding operations.

Animal fats and oils tend to develop disagreeable odors due to oxidative and bacterial rancidity. However, this problem can be reduced by corrective chemical processing and the use of germicidal additives.

Vegetable. Vegetable oils and fats used as polar additives are obtained by crushing and rendering the kernels, seeds, and whole fruit of specific plants. The resulting oils are liquid and contain a certain percentage of unsaturated fat.

Vegetable oils are divided into two categories: *drying* and *nondrying* oils. The drying type form a tough elastic film when exposed to the atmosphere due to oxygen absorption. The drying characteristics of some oils, however, is due to the high percentage of unsaturated fatty acids which they contain.

The nondrying vegetable oils contain the least amount of unsaturated fats and will *not* form the tough elastic film when exposed to the atmosphere.

Because of their "drying properties" and tendency to gum, the unsaturated, drying type vegetable oils must be further processed before they can be used as lubricants, or to make soaps in cutting fluids. The nondrying type vegetable oils, such as palm oil, coconut oil, and castor oil, because of their low degree of unsaturation, do not need further processing. They are used extensively in the manufacture of cutting fluids and other lubricants.

Marine. The polar additives of marine origin (fatty tissue of fish, whales, and other marine animals) are liquids which exhibit a high degree of unsaturation. The one notable exception to this is sperm oil. This oil is a fatty lubricant derived from the sperm whale and has a degree of unsaturation similar to many animal fatty oils. Sperm oil was one of the most popular "lubricants" of marine origin used in cutting fluids. However, it is actually a liquid wax rather than a fatty oil. It is an excellent boundary lubricant and is highly resistant to gumming and viscosity decrease when subjected to high temperatures and pressures.

The marine oils are good boundary lubricants and offer a handling advantage because of their liquid form. However, they all have a characteristic "fishy" odor and must be treated to reduce or eliminate this odor before they are industrially acceptable.

Chemical Additives. To produce acceptable parts economically from difficult-to-machine metals, the polar additives used in cutting oils must be supplemented with more effective boundary or extreme pressure lubricants. They provide tougher, more stable solid film lubrication. This supplementary extreme pressure lubrication is furnished by adding to the carrier substances which contain reactive sulfur, chlorine, or phosphorus components.

When subjected to the temperatures found at the chip/tool interface, the sulfur-, chlorine-, or sulfur- and chlorine-bearing extreme pressure lubricant apparently reacts to effect: (1) lower cutting forces, (2) reduction of chip thickness, and (3) improved finish.

Prior to extensive research devoted to determining the exact mechanism of reaction, general opinion was that a low-shear-strength metallic sulfide or chloride film was formed at the chip/tool interface and tool flank/cut surface interface so that lowered friction reduced the tendency toward formation of the built-up edge. Today, however, greater acceptance is being given to the theory that the sulfur, chlorine, or phosphorus may react favorably, not only on the surface of the chip, but internally to a depth sufficient to reduce forces in the shear zone. The combined effects of chemical reactions (both internal and external) will markedly reduce forces in both areas. The total energy reduction can be as much as 50 to 80%. Hence, extreme pressure lubricants can help control the built-up edge, improve finishes, and help to control tool life.

Sulfur. Sulfur may be added to the cutting oil in the form of a sulfurized mineral oil or a sulfurized fat. Sulfurized mineral oils are prepared by simply dissolving elemental sulfur in hot mineral oil. Sulfurization of fats requires much higher temperatures than sulfurization of mineral oils. The latter is a matter of dissolution, while combining sulfur with a fatty oil is an exothermic, chemical reaction. This results in a stronger chemical bond between the fatty oil and the sulfur atom.

More energy is required to sulfurize a fat than to dissolve sulfur in mineral oil. As a result, more energy is required to break the sulfur away from the fatty oil to react at the chip/tool interface during the cutting operation. For this reason, the sulfurized mineral oil is more "active" at lower temperatures than the sulfurized fatty oil. However, since the temperatures at the chip/tool interface exceed those required to produce either a sulfurized mineral oil or a sulfurized fat, the difference is of little consequence in relation to their comparative effectiveness.

The staining tendencies of cutting oils containing sulfurized extreme pressure lubricants vary widely. Because the sulfur dissolved in a mineral oil is very loosely held, it is very reactive at low temperatures and tends to stain severely nonferrous metals such as copper, brass, and bronze. In some cases it will also stain a freshly machined steel surface. In comparison, the fatty oil which has been sulfurized at higher temperatures will not release active sulfur as readily and, therefore, has less tendency to stain nonferrous metals or steel.

It is possible to manufacture sulfurized fatty materials that will not stain any metals under almost any conditions encountered in machining and grinding operations. However, a completely nonstaining oil is not a universal requirement. On many operations, the more active oils will perform satisfactorily and, in many cases, at lower cost.

Chlorine. Chlorine also functions as an effective extreme pressure lubricant in cutting oils. It is generally incorporated in the carrier mineral oil as a long-chain chlorinated wax, or as a high-molecular weight chlorinated ester.

The chlorine additive is sometimes introduced into cutting oils in combination with sulfur as a sulfo-chlorinated mineral oil or as sulfo-chlorinated fatty oil. This type of extreme-pressure-bearing component is manufactured by reacting a molecule, containing both sulfur and chlorine at high temperature, with a carrier wax, mineral oil, or fatty oil. Chlorine reacts and functions at the chip/tool interface and the tool flank/cut surface interface in essentially the same manner as sulfur. However, it is more reactive than sulfur and begins to combine with the substrate metal at lower temperatures (see Figure 2-4). Usually, chlorine-bearing extreme pressure lubricants will not stain most nonferrous metals.

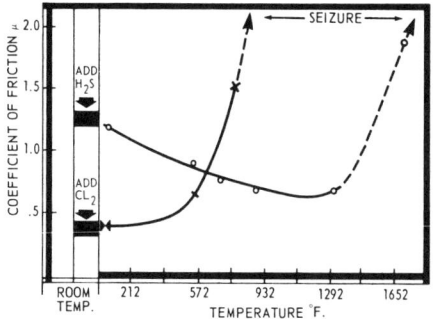

Figure 2-4. *Relative effectiveness of chlorine and sulfur in reducing friction at various temperatures.*

Because of their high chemical reactivity, chlorine-bearing extreme pressure lubricants are always compounded with inhibiting or neutralizing ingredients to prevent corrosion of ferrous surfaces. When chlorine first came into use as an extreme pressure lubricant, corrosion of ferrous surfaces was a common problem. Today, with the use of inhibitors and better manufacturing techniques, chlorine-bearing extreme pressure lubricants are much more stable. Chemical reaction between the chlorine and the metal is restricted to the chip/tool interface where the temperatures are high, eliminating the corrosion problems created by the acidic byproducts given off by unstable, uninhibited chlorine-bearing materials.

Phosphorus. When added to the carrier mineral oil as an organic phosphate or metallic phosphate, phosphorus will perform as a mild extreme pressure lubricant or antifriction additive. It is not as effective in preventing metal welding as sulfur and chlorine because the phosphide film will break down at lower temperatures. Phosphorus is most effective in reducing friction and wear. In addition, most phosphorus-bearing extreme pressure lubricants are nonstaining to most ferrous and nonferrous metals.

Compounded Cutting Oils. Cutting oil manufacturers sometimes use a combination of the polar and chemical additives dispersed in a carrier mineral oil. The balance or combination of additives used depends upon the requirements of the cutting operation and metal being processed.

A partial list of compounded cutting oils available includes:

1. Fatty mineral oil blends
2. Sulfurized fatty mineral oils
3. Chlorinated fatty or nonfatty mineral oils
4. Sulfurized mineral oils
5. Sulfo-chlorinated mineral oils
6. Sulfo-chlorinated fatty oil blends
7. Any of the above also containing phosphorus, other metal salts, and solid lubricants
8. Any combination of any of the above

Solid Lubricants. The solid lubricant type of extreme pressure additive is used to a very limited extent in cutting fluids, especially if the fluid is being recirculated. To be effective, the solid lubricant, which is dispersed in the carrier oil, must remain uniformly suspended. Constant uniform suspension is difficult to achieve; therefore, the solid lubricant particle may precipitate and cause line clogging and dirty machine tool surfaces.

Molybdenum disulfide is an example of a common solid lubricant. It has a flat laminar structure which adheres to metal surfaces to form a coating with good friction reducing properties. In addition, under the heat and pressure of the cutting operation, the sulfur held in the disulfide molecule combines to form a low shear-strength, extreme pressure metallic sulfide film.

EMULSIFIED OILS (WATER MISCIBLE)

An emulsion is a suspension of oil droplets in water made by blending the oil with emulsifying agents and other materials. The addition of animal or vegetable fats or oils, or other esters, produces "super-fatted" emulsions of greater lubricating value. The addition of sulfur, chlorine or phosphorus products produces fluids of yet greater lubricating value which are extreme pressure emulsions.

Water miscible fluids form mixtures ranging from emulsions to solutions when mixed with water. Due to its high specific heat, high thermal conductivity, and high heat of vaporization, water is one of the most effective cooling media known. Blended with water, water miscible fluids provide the combined cooling and lubrication required by metal removal operations conducted at high speeds and lower pressures.

Water miscible fluids are available in many forms and variations. Distinctions between them are usually made on the basis of appearance and performance. When mixed with water, however, they can be classified on an appearance basis as *normal emulsions* or *chemical fluids*. The latter is discussed in the following section.

The normal water miscible emulsion discussed in this section contains emulsified particles large enough to reflect almost all light and appears opaque or "milky." Figure 2-5 classifies the different water miscible products according to ideal particle size.

Water miscible cutting fluids offer the following advantages:

1. Reduction of heat--allowing higher cutting speeds
2. Cleaner conditions
3. More economical--dilution with water brings application costs down
4. Better operator acceptance--cooler, cleaner parts
5. Improved health and safety benefits--no fire hazard; reduction of oil misting and fogging

	DECREASING PARTICLE SIZE	
NORMAL EMULSION	SEMI-CHEMICAL	CHEMICAL
PARTICLE SIZE LARGER THAN 0.000004 IN.	PARTICLE SIZE BETWEEN 0.000004 IN. AND 0.00000004 IN.	PARTICLE SIZE SMALLER THAN 0.00000004 IN.
MOST NORMAL CUTTING FLUID EMULSIONS RANGE FROM: 0.0002 IN. TO 0.00008 IN.	COLLOID SOLUTION PARTICLES	TRUE SOLUTION MOLECULAR AND IONIC DISPERSOIDS
PARTICLES CAN BE SEEN UNDER ORDINARY MICROSCOPE	PARTICLES CAN BE SEEN UNDER ELECTRON MICROSCOPE	PARTICLES CANNOT BE SEEN UNDER MICROSCOPE OF ANY KIND

Figure 2-5. *Water miscible products classified according to ideal particle size.*

Water miscible fluids provide less rust control and significantly decreased corrosion inhibition due to the introduction of water. In recent years, however, significant strides have been made with the use of new, more effective corrosion preventive additives. The need for "in-process corrosion preventives" has been reduced drastically, and several days to a week of corrosion inhibition can be expected on many applications.

Because the viscosity of a water miscible fluid mixture is almost equal to that of water, its inherent film strength and lubrication properties are inferior to those of most straight cutting oils. The inherent viscosity of a straight cutting oil supplies "hydraulic cushioning" to reduce shock (especially on an abrupt cutting operation such as broaching) and reduces abrasive wear. This shortcoming in water miscible fluids is also apparent in severe grinding operations, such as form, thread, and crush grinding, where wheel form must be maintained.

The soaps, wetting agents, and couplers used as emulsifiers in water miscible fluids reduce surface tension significantly. With this reduction in surface tension, a liquid becomes more prone to foaming when subjected to shear and turbulence. For this reason, water miscible fluids sometimes cause a foaming problem in operations such as gundrilling, flat-bed and double-disc grinding. However, with the use of special wetting agents and foam depressants, water miscible fluids can be rendered sufficiently non-foaming in almost all operations to be effective.

Emulsified Mineral Oils

The common emulsified mineral oils contains a paraffinic or naphthenic mineral oil ranging in viscosity from 100 to 500 secs. at 100° F (37.8° C). The emulsifier may consist of petroleum sulfonates, amine soaps, rosin soaps, naphthenic acids, etc. Popular "coupling" agents that assure stability in the base form are complex alcohols and nonionic wetting agents. Other organic polar materials may be added to increase corrosion inhibition. Microbiocides are often included to resist attack by bacteria, fungi, or mold, and to extend emulsion life. Normal use concentrations may range from 2 to 10%.

Highly-Fatted Water Miscible Fluids

Fatty oils and fatty acids such as sperm oils, lard oil, esters, and other mixtures, are often added to emulsion concentrates to increase lubrication. This type of product is often used on soft, stringy nonferrous alloys.

Extreme Pressure Emulsifiable Oils

Sulfur, chlorine, and phosphorus serve the same function in water miscible products as in cutting oils. Generally, water miscible cutting fluids containing extreme pressure additives are termed heavyduty solubles and are capable, in some cases, of replacing cutting oils.

Fatty and extreme-pressure-bearing water miscible fluids are used in rather rich concentrations, such as 1 part oil to 5 to 15 parts water.

CHEMICAL AND SEMICHEMICAL FLUIDS (WATER MISCIBLE)

The clarity of the chemical and semichemical solutions, when diluted with water, varies from translucent to completely clear. This is in contrast with the "milky" or opaque emulsion formed by the normal mineral-oil-based water miscible fluids. The chemical fluid is completely clear because its particle size as shown in Figure 2-6 is small enough to transmit almost all incident light. The solutions of intermediate particle size are classified

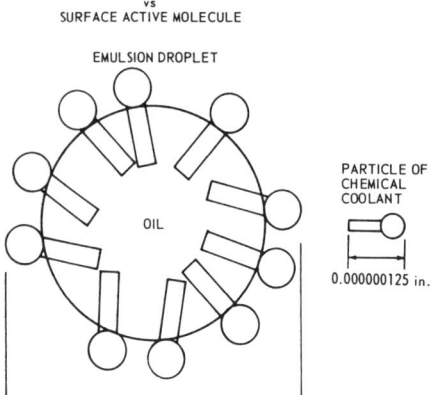

Figure 2-6. *Relative size of surface active molecule compared with typical emulsion droplets in a "soluble oil" mixture. (Courtesy, Master Chemical Corporation)*

as semichemical; these products form hazy or slightly translucent emulsions.

Chemical Fluids

The chemical fluids are classified into three primary types: (1) true solution type, (2) wetting agent type, and (3) wetting agent type with extreme pressure lubricant. Those which form a clear, transparent solution are water-based. They may contain inorganic materials such as borates, molybdates, and phosphates for corrosion inhibition, or organic materials such as amines and amides for supplementary protection. Complex alcohols are usually used as humectants. Surface active agents are added to reduce surface and interfacial tension and to promote wetting and lubrication.

The many such chemical products now available fall into two general classes: those *with*, and those *without*, wetting agents and good lubricity. Those *with* lubricity and wetting agents have low surface tension, good rust-inhibiting properties, and leave a liquid residue. In normal operations, their slight tendency to foam is not a factor. Fluids may foam excessively when air is beaten into them, in disc-type surface grinders, for example. The excellent lubricating qualities of chemical fluids allow machine slides, turrets, and other moving parts to function smoothly--an important factor with increasing automation.

The group of chemical fluids *without* wetting agents usually do not have much lubricity, and many leave crystalline or wax like deposits upon evaporation of the water. This eventually interferes with machine action. Many do not have enough lubricity to do the "tough" jobs. These fluids do not produce as good a finish in machining and grinding operations as products with good lubricity characteristics. Chemical fluids without wetting agents usually do not "scum" in hard water.

Semichemical Fluids

Semichemical fluids, unlike chemical fluids, contain a small amount of mineral oil plus additives to further enhance lubrication properties. The semichemicals are gaining favor in industry today. They incorporate the best qualities of both chemicals and normal water emulsions. Both chemical and semichemical fluids are now available containing chlorine, sulfur, or phosphorus additives which afford extreme pressure or boundary lubrication effects. Because of these additives, chemical or semichemical fluids can be used on some of the more difficult machining and grinding applications. A concentration range may vary from 2 to 10%.

Advantages and Disadvantages

In general, the chemical and semichemical fluids offer the following advantages:
1. Rapid heat dissipation and good size control
2. A high degree of cleanliness resulting in clean machine-tool surfaces and clean coolant troughs
3. Good detergent properties which aid in the maintenance of open and free-cutting grinding wheels
4. Excellent workpiece visibility
5. Very light residual film which is easy to remove
6. Easy mixing with very little agitation necessary
7. Excellent resistance to rancidity and therefore good tank life
8. Relatively easy concentration control with no interference from tramp oils

The chemical and semichemical fluids both contain considerably less mineral oil than the normal emulsifiable or water miscible fluids. (The "true" synthetic or chemical fluid contains *no* mineral oil.) In addition, emulsion particle size is much smaller as was shown in Figure 2-5. The combination of these two factors provides the chemical and semichemical fluids with much better long-term stability than the normal emulsion.

Spoiled normal emulsions are partially the result of anaerobic bacteria which proliferate rapidly under an oil layer or oil "seal" formed by a split emulsion or tramp oil accumulation. Emulsions containing high concentrations of fatty lubricants and emulsifiers upon which this type of organism subsists are also more susceptible. Generally, chemicals and semichemicals contain little or no mineral oil or other types of lubricants upon which anaerobic bacteria feed. This characteristic, combined with their ability to remain stable, generally gives them better tank life than the normal emulsion.

Chemical and semichemical fluids can be formulated to provide good wet "contact" corrosion control with the use of organic and inorganic inhibitors. Products containing these inhibitors also exhibit relatively good protection from overall atmospheric corrosion. A normal water miscible fluid containing organic corrosion inhibitors, polar lubricants, and a high percentage of mineral oil, will give superior overall corrosion protection over straight chemical fluids. The fatty lubricants and mineral oil form a "barrier" film on the workpiece which provides an extra measure of protection.

On the negative side, these disadvantages are sometimes encountered with chemical and semichemical fluids:

1. Lack of lubrication "oiliness" which may cause sticking in the moving parts of machine tools
2. High detergency may defat and irritate sensitive hands where operator exposure is continual for long periods of time
3. In comparison to oils, less rust control and significantly decreased corrosion inhibition, inferior inherent film strength and lubrication properties, and some tendency to foam.

GASEOUS FLUIDS

Gaseous fluids can perform both cooling and lubricating functions. Some gases will oxidize the newly-formed and chemically-clean chip, thereby preventing welding to the tool and reducing friction. All gases provide cooling either by convection, or when chilled to liquid form, by vaporization.

Air

Air is the most commonly used gaseous cutting fluid. It is the sole fluid constituent in "dry" cutting and is also present, of course, when liquid fluids are used. The cooling and lubricating action of air is taken for granted because it is always present.

Air can also be used as a compressed gas to provide better cooling. A stream of compressed "shop air" directed at the cutting zone removes more heat by forced convection than would occur by natural convection. In addition, compressed air can be used to blow chips away.

Gases for Special Applications

Other gases have been used as cutting fluids. Their high cost generally makes them uneconomical in production except in very special applications.

Inert and semi-inert gases, such as argon, helium, and nitrogen, have been used to prevent the oxidation of workpiece and chip where unusual chemical properties are required. The cutting area is flooded with the gas to exclude oxygen. The gas must be applied in a sealed container (such as a welding dry box) so that the cutter does not blow the gas away.

Gases with boiling points below room temperature can be used as coolants. Freon or carbon dioxide can be compressed and sprayed at the cutting zone to give evaporative cooling to temperatures well below $0°$ F ($17.8°$ C). The use of liquid argon or nitrogen allows cooling to several hundred degrees below $0°F$. Increases in tool life can be expected from all of these gaseous coolants, but their cost will be extremely high.

MISCELLANEOUS FLUIDS

In smaller shops where safety and industrial hygiene rules are not stringently enforced, limited application of carbon tetrachloride and trichloromethane will be found. These are highly volatile, short-chained hydrocarbons containing a high concentration of chlorine. These materials wet well, dissipate heat rapidly and, being relatively unstable, provide a metallic chloride shear-strength reducing effect at low temperatures. For these reasons, they do facilitate severe machining operations on difficult metals such as tapping and threading of 300 Series stainless steels and other heat-resistant alloys. They are toxic if inhaled or absorbed through the skin and will cause skin defatting. Another disadvantage is possible stress corrosion cracking or failure induced by components containing active chloride ions. These fluids, although promoting good chip curl, cause tools to wear rapidly with loss of size control, particularly in tapping operations.

REFERENCES

1. F. W. Taylor, *On the Art of Cutting Metals*, *ASME*, New York, New York, 1906.

CHAPTER 3
HOW CUTTING FLUIDS ARE APPLIED

The proper physical delivery of cutting fluids to the point of cut is one of the most neglected aspects of cutting fluid application. Yet, it is one of the most important. Unless the fluid is carefully placed, it cannot perform its function. Unfortunately, placement is often left to the operator who, not understanding the fluid's functions, places it for his personal convenience rather than for maximum effectiveness.

When a fluid is chosen for its lubrication qualities, it must be directed so it can form a film between the tool, the work, and the chip. A good lubricating fluid dribbled haphazardly onto the workpiece prevents rust and lubricates the ways, but does not improve the cutting process. To obtain improved finish and tool life, the fluid must be directed to the place where these benefits occur--the cutting edge of the tool.

Similarly, when a fluid is chosen for its cooling properties, it must be directed to where cooling is required--at the cutting edge of the tool. While a general flow over the workpiece does help to cool the work, it is only by forcing the fluid *into* the cutting area that heat can be removed as it is generated.

A common example of the misapplication of a coolant is in surface grinding, where the fluid flows on the workpiece. Here, the grinding wheel acts as a fan, blowing the fluid away and preventing it from flowing *in* and cooling the grinding process.

A secondary advantage of good fluid distribution design is the removal of chips. Proper placement of nozzles can prevent blockage or packing of the chip in the flutes of milling cutters and tool breakage. It can reduce or eliminate the need for brushing or blowing the chips away by hand labor, and remove the still warm chips from the working areas of the machine.

The tendency of hard and brittle cutting tool materials (such as carbide and ceramic) to crack when subjected to thermal shock can be controlled by proper design of the distribution system. Cyclic heating and cooling of the edge produces microcracks in the surface which grow until small pieces of the edge break away. As the edge continues to crumble, premature failure occurs. However, properly placed nozzles, a ring distributor, or a mist apparatus can give continuous gentle cooling. Where possible, the operator should not continuously start and stop the flow to "see what is happening" because the alternate heating and chilling promotes the type of failure described above.

Many methods exist for applying cutting fluid to the tool/work interface. Each has its particular applications. To select the best approach, examine the advantages, disadvantages, and economics associated with the specific job to be done.

MANUAL APPLICATION

Manual application of cutting fluids is the most simple and least costly of the various methods in use. It consists of an oil can or a container filled with fluid, and a paint brush for applying it.

Drilling and Tapping

Where a small number of holes must be drilled or tapped on a machine not equipped with a pumping system, manual application of the cutting fluid is an effective method. A fairly continuous stream of fluid from an oil can directed at a drill that is retracted often, reaches the area where cutting takes place to cool, lubricate, and help remove chips. Similarly, a lubricant brushed onto a tap is effective in reducing forces and, therefore, the number of broken taps. The manual application of solid or semisolid lubricants can also be used efficiently in tapping. Wax is inserted into a blind hole so the tap forces it out, carrying the chips with it. Since it is forced into the cutting zone, the wax also performs an effective lubrication function.

Manual application is sometimes used in conjunction with more sophisticated systems when two different operations are performed on the same machine. For example, the flood cooling system of a drilling machine could be filled with a moderately active fluid for the drilling operation, and a highly active fluid applied manually for a subsequent tapping operation. Many active sulfurized and chlorinated cutting fluids will, however, hydrolyze in the presence of water to release sulfurous and hydrochloric acids which can shorten the life of the coolant and cause corrosion on machines and workpieces, and even skin irritation on machine operators.

For this reason, manually applied tapping fluids which are chemically stable in the presence of water should be selected. In some cases, it may be possible to substitute the coolant concentrate for a special tapping oil and eliminate this source of contamination of the cutting fluids. Figure 3-1 shows an automatic tapping fluid applicator designed for installation on NC/CNC machining centers to eliminate interrupting the machining cycle for manual application of tapping fluid.

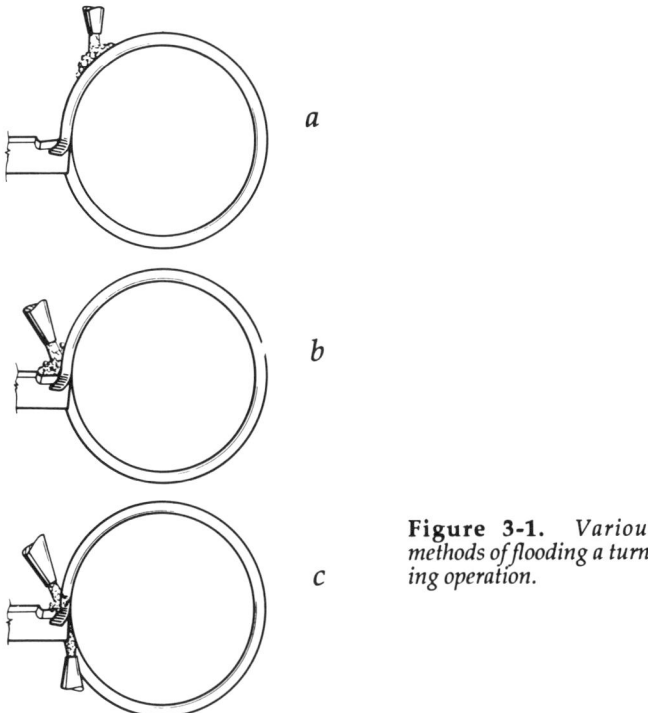

Figure 3-1. *Various methods of flooding a turning operation.*

Turning and Milling

Manual application is often used in turning or milling but is generally ineffective. The operator who applies oil with a brush on a small turning or milling job can only reach the wrong side of the chip and never reaches the *area of the cut*. In addition, the effect of manual application by an oil can is hampered by inconsistency. One moment the cutting area is well flooded, while the intermittent moment between spurts forces a dry cut, i.e., there may be no fluid in the cutting zone 50 to 75% of the time. In certain instances, however, the advantages of manual application may outweigh the disadvantages of the necessity to clean the part and machine when the operation is over.

FLOOD APPLICATION

Flooding is the most common fluid application system used, and also one of the most commonly abused. A low-pressure pump delivers the cutting fluid through pipes and valves to a nozzle situated over the cutting zone through which it flows down and floods the tool, work, and chip. The fluid is then collected in the chip pan and returned by gravity to the coolant sump.

Though the system is simple and commonly used, it must be carefully designed for successful application. The fluid *must* be delivered to the *exact* spot to be effective. The following illustrations demonstrate effective designs and techniques which can be applied to particular operations.

Turning

Figure 3-1 shows various methods of flooding a turning operation. In Figure 3-1*a*, the fluid is not directed at the point of cut; therefore, little fluid actually gets to that point. Most is thrown off by centrifugal force. This situation is contrasted with the setup in Figure 3-1*b*. The fluid is pointed directly at the spot where the chip is forming. When fluid flows over this area, good cooling is achieved.

The two nozzles in Figure 3-1*c* provide even better lubricating and cooling efficiency. The flow from the top nozzle cannot reach the rake face of the tool easily, since the chip covers it completely. However, the flow from the lower nozzle is not similarly hindered and can be forced between the work and tool to help lubricate at low cutting speeds.

It is questionable if any lubrication is provided by a cutting fluid at extremely high cutting speeds. The high velocities of work and chip, coupled with high unit pressures, do not allow penetration of the fluid or time for chemical reactions. This is the reason that fluids for high speed cutting are chosen primarily for their cooling ability.

Milling

Figures 3-2 to 3-4 demonstrate the same principle described above, but applied to milling operations. Figure 3-2 shows the use of two nozzles--one on each side of the cutter to ensure thorough flooding of the cutting zone. An even better design is shown in Figure 3-3 where fluid from one nozzle is pumped *through* the cutting zone by the cutter teeth while the other nozzle washes the chips from the cutter as they emerge. Standard round nozzles can be used for narrow cutters. The fan-shaped nozzles required

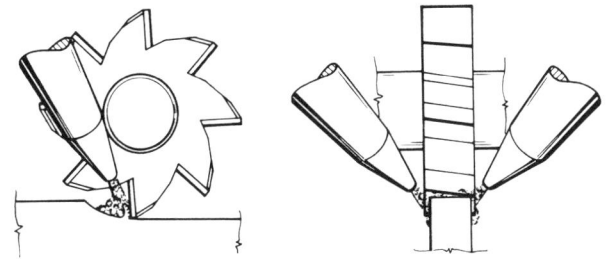

Figure 3-2. *Use of two nozzles to ensure flooding of the cutting zone.*

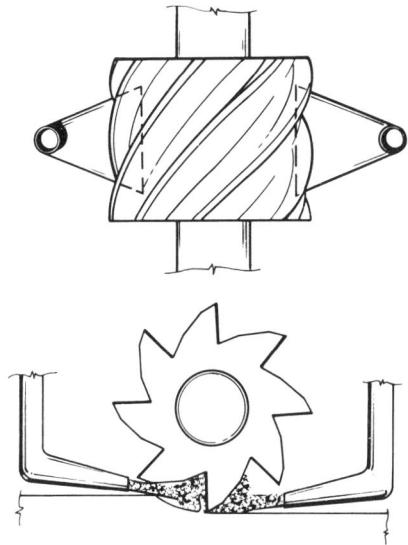

Figure 3-3. *Fluid is pumped from right nozzle by cutter teeth while fluid from left nozzle washes away chips as they emerge.*

for wider cutters can be easily made in a variety of sizes to fit. The width of the fan should be at least 3/4 of the width of cut to provide good coverage.

Figure 3-4 shows a ring distributor used with a face mill. It consists of a tube with many small holes which direct the fluid against the entire circumference of the cutter to give even cooling. This is the most effective semiuniversal method of applying fluid to this difficult-to-cool-and-lubricate operation. If a particular size face mill is used often, the ring distributor can be supplemented with a special fan nozzle which has a curved opening to match the cutter radius.

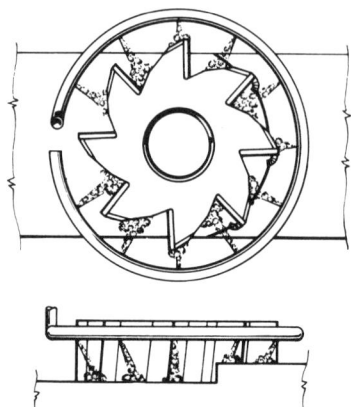

Figure 3-4. *Ring distributor used with face mill directs fluid against the entire circumference of the cutter for even cooling.*

Drilling

There is no really good way to cool and lubricate the cutting area when drilling with a common twist drill. Very little fluid gets down to the zone of cutting. Oil hole drills, when they can be used, will solve this problem. An oil hole drill has single or multiple holes running down through the body of the drill for fluid passage. The fluid, under pressure, is transferred to the drill from the reservoir by a rotating gland and is forced into the cutting zone. The oil flowing from the hole helps to force chips out as well.

Gundrilling

An arrangement similar to the above is used on gundrills, except that pressures up to 1000 psi (6895 kPa) are used to drive the chips from the deep holes. Whenever these high pressures are used, plastic shields must be provided to protect the operator and the surrounding areas.

BTA and Ejector Drilling

In BTA or ejector drilling, the fluid is injected to the point of cut through the tool itself. The fluid and chips are then flushed through flutes on the outside surface of the drill or an internal channel within the drill center. It is important to have a high-quality, light viscosity fluid that lets the chips slide freely off the cutting edge and flushes them along the discharge flutes or channels. Workpiece material limits the use of these type of drills because it must produce chips that will turn ahead of the tool under pressure from the fluid so the fluid will not clog the fluid supply channels and stop lubrication to the tip.

Grinding

The normal methods of applying fluids to grinding remove little cutting heat until after it has dissipated into the mass of the workpiece. Even fluid applied to the work in the vicinity of the wheel does not reach the cutting zone. This is because the grinding wheel, rotating at 6500 sfm (1981 m/min) or more, picks up a film of air which encloses it and excludes the fluid. Two of the more common methods of penetrating this air film are: (1) passing the fluid through the wheel and (2) the use of special nozzles.

Passing Fluid Through the Wheel. This method requires a hollow grinding spindle end and a gland seal for transferring the fluid to the revolving spindle. The fluid is introduced into the center of the wheel and passes through the fine porous passages to the circumference of the wheel where it is thrown off by centrifugal force. The difficulties encountered in this system, other than the special equipment requirements, are twofold. First, the passages in the wheel are very small and any particles of swarf can easily plug them. The fluid must be filtered through a 3 micron (.0001 inch) filter. Second, the fluid is thrown off over the entire circumference even though only a small portion of the wheel is in contact with the workpiece at any one time. This creates a mist of fluid over the entire operation which must be contained.

Special Nozzles for Grinding. Like the "through the wheel" technique, special nozzles for grinding were initially developed for electrolytically-assisted (EAG) grinding. EAG requires that the area between the special grinding wheel and the workpiece be completely flooded with electrolyte for the process to work. Some of these nozzles are simply fan-shaped nozzles pressed against the wheel so the wheel grinds the nozzle to its exact contour. The edges break the air film and allow the wheel to carry the fluid into the cutting zone. The nozzle must be kept as close as possible to the workpiece so all of the fluid is not thrown off before the wheel reaches the cut.

A more sophisticated nozzle, designed specifically for conventional grinding, is shown in Figure 3-5. This nozzle will accommodate a range of wheel sizes as long as the lip is adjusted against the wheel. The nozzle must be at least 3/4 of the width of the wheel, but should not overlap the edges of the wheel at any point. All special nozzles will throw some mist.

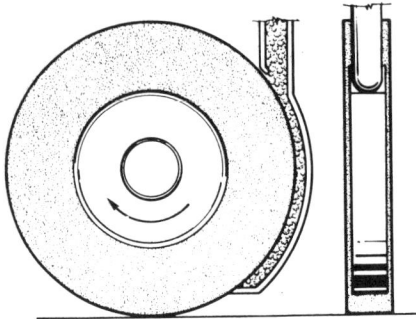

Figure 3-5. *Special nozzle developed for grinding operations will accommodate a range of wheel sizes.*

Creep Feed Grinding

Creep feed grinding involves high energy requirements and narrow specifications that place severe demands on coolants and coolant systems. The fluids must provide cooling, cleaning, lubrication and rust protection. A great deal of heat is produced in the grinding process which must be removed from the workpiece to maintain size control. This requires larger system capacity to maintain temperature control of the fluid as well as to allow settling time for the fine swarf. If swarf is carried back to the workpiece, it results in poor surface finish and/or loading of the grinding wheel, resulting in reduced cutting effectiveness. As a rule of thumb, the settling tank should have a minimum capacity of five times the coolant flow rate per minute to accommodate temperature, settling and foam. Therefore, a system pumping 120 gallons per minute (454 L) would need at least a 600 gallon tank (2271 L). Filters and chillers can also be used to further control grit, temperature and foam. Again, fluid must also be applied correctly to the area of contact. This is critical because the rate of flow and pressure of systems increase the need for greater precision in nozzle shape and position. Also, the high RPM's of the grinding wheel prevent it from reaching the contact point. Special air scraper and other types of nozzles are available to counter this problem.

AIR-CARRIED MIST

Normally, fluids for mist application are chosen for their ability to cool rather than lubricate. The small volume of liquid reaching the cutting zone precludes the formation of hydrodynamic lubrication films. However, the use of fluids with extreme pressure additives may provide some boundary lubrication. This does not mean that oil-base fluids are never used. Since the heat of vaporization of oils is only 1/7 that of water, and the major beneficial effect is cooling by vaporization, water-base fluids are generally preferred. An exception is the use of oil in chip breaker grinding with diamond wheels.

Advantages and Disadvantages

Special hazards are created when using air-carried mist which must be considered from the standpoint of both the operator's health and comfort. The liquids used are the same as those used in conventional fluid systems. However, they are in finely-divided particles which can be inhaled. Conventional cutting fluids are not highly toxic but certain individuals might be affected by them over a period of time. In addition, the feel or smell of the confined mist, vapor, and smoke from the hot workpiece may be offensive to some operators. Therefore, it is necessary that the operator be protected from contact with the mist. One simple way is to never direct the mist toward the operator or an operator on a nearby machine. Mist collectors which draw air from the working area and filter it before returning it to the room are very effective when the inlet nozzle can be placed close to the point of cut. However, they require periodic cleaning to ensure that good suction is maintained. Ordinary fans will do an adequate job when the mist unit is mounted only temporarily and the mist can be blown in a safe direction. This setup will not be satisfactory in a crowded shop having several units unless extremely good natural ventilation is available.

A possibility of fire exists with a mist of volatile and combustible fluids. However, fire becomes only a remote chance if good ventilation and conventional water-base fluids are used, since the heat is removed by vaporizing the small amount of fluid. Smoking can sometimes be reduced by increasing the flow of fluid, but this method approaches flood cooling and does not take full advantage of the considerable cooling effect of vaporization.

Mist cooling does not require the splash guards, chip pans, and return hoses associated with other cutting fluid application techniques. Only a small amount of liquid is used and it generally dries on the part or can be wiped away easily. Therefore, this system is particularly suitable for fitting to machine tools designed without proper provisions for cutting fluids and also makes the use of fluids practical for use with portable tools.

All equipment for applying various types of mist cutting fluid operates on one basic principle. A small quantity of liquid coolant is introduced into a high-velocity stream of air to form a mist. This stream is then directed at, or into, the cutting zone. Since the liquid is dispersed into fine particles, much of it evaporates on contact with the hot tool, workpiece, or chip. Since it takes 100 times more heat to evaporate a quantity of water than to raise its temperature 10°F (-12.2°C), the small amount of coolant which is evaporated can do an effective job--particularly since the high-velocity airstream can carry it to the spot where it is most needed.

Aspirator and Pressure-Fed Generators

Two types of mist generators are normally used: aspirator or pressure-fed. The aspirator type works on the same principle as a household fly sprayer. A stream of shop air is blown over the open end of a tube which is immersed in the coolant. This creates a partial vacuum. Fluid is drawn up to the top of the tube where the airstream picks it up and carries it to the workpiece.

The pressure-fed type uses positive pressure rather than a vacuum to feed the coolant into the airstream. The coolant is held in an air-tight tank under a pressure of 10-40 psi (69-276 kPa) which forces the coolant out into the airstream.

The aspirator type of equipment is generally much less expensive but has certain limitations. Since only a weak vacuum is formed, the aspirating nozzles and control valves must be located on the fluid sump. The hose to the spray nozzle must be short (3 to 4 feet/0.9 to 1.2 m) so that the mist will not have time to settle out. The short hose, coupled with the need for the operator to adjust or turn off the mist with the aspirator valves, limits the placement of the tank. The pressure-fed type mixes the mist at the spray nozzle from two hoses, one containing air and one containing the coolant. The control valve is also mounted on the nozzle, allowing easy access by the operator. Several nozzles can be operated from one pressure system, either on one machine for better coverage or on adjacent machines.

Portable Systems

Portable mist systems are also available which use bottled high-pressure gas in place of shop air lines. Suitable pressure-reducing valves must be incorporated, and other safety precautions pertaining to high-pressure gas bottles must be heeded.

End Milling

Mist cooling can be used on almost any operation, but seems best suited to cuts where the speed is high and the areas of cut are low. End milling is perhaps the most common application. One or more nozzles are placed so the mist stream enters the cutting zone without being obstructed by the workpiece configuration. Since the mist will carry several inches, this can be accomplished so the nozzle is not bumped or deflected by the work, even in three-dimensional cutting.

Similar setups can be used in other types of milling operations. When a wide milling cut is encountered, a fan nozzle which produces a long narrow spray pattern can be used. Since spray is carried in the air, it is ideally suited for application from a position under the tool when turning. However, an automatic shut off must be provided so that the operator does not inhale the mist when handling the part.

Drilling

In drilling, mist cooling is most successful for use with shallow holes or where an oil hole drill is available. In holes deeper than one diameter, the mist does not penetrate to give cooling at the point of cut unless an oil hole type drill is used.

N/C and Transfer Machines

These machines frequently use a water miscible type of fluid applied as a flood. (Certain individual operations, however, such as reaming, tapping, and threading, can be accomplished better by a more lubricating type of fluid.) Water miscible fluids can be applied as a drip, mist, or intermittent flood by actuating solenoid valves to use the minimum amounts of fluid, in the proper sequence, without excessively contaminating the main body of the fluid. This results in optimum fluid usage and best machine operation.

CHILLED FLUID APPLICATION

Many attempts have been made to apply fluids chilled to temperatures below ambient. They have generally resulted in benefits in terms of tool life, but monetary savings have been insufficient to justify the cost of cooling the fluid. However, it is sometimes worthwhile to try this method in cases where nothing else will work. Table 3-1 lists some of the fluids that have been tried and the temperatures at which they can be used.

Fluids can be cooled by various methods depending on the sophistication desired. Liquid gases such as argon or nitrogen are purchased in the liquid state in large Dewar flasks. Liquids with boiling points above room temperature can be chilled by mechanical refrigeration units or by filling the sump with dry ice. The latter method can reach temperatures of -110°F (-79°C) with little initial expense.

Table 3-1. Fluids for Use at Subambient Temperatures

Fluid	Temperature Used
75% Stoddard Solvent* 25% Trichlorethylene	−76°F.
50% Stoddard Solvent 50% Trichlorethylene	−98°F.
Liquid Nitrogen	−320°F.
Liquid Argon	−301°F.
CO_2 Mist	−84°F.

*500 lbs. of dry ice will maintain this temperature in a 140 gal. system for an 8 hr. shift. 400 lbs. are added at the beginning and 100 lbs. 5 hrs. later.

Coolant pumps must be designed for the temperature expected. Special low-temperature pumps have special seals and bearings located some distance from the cold fluid so the pump lubricant does not congeal. The machine tool, sump, and piping--to both the work area and return--must be insulated to prevent heat transfer. The only heat that should be absorbed by the fluid is that given off at the cutter. Cooling the fluid is expensive so no unnecessary heat should be allowed to warm it. Any valves in the piping system must be specially designed and be capable of operating at the low temperatures encountered.

The operator must be protected so the temperatures used will not cause immediate freezing of flesh. Insulated nonabsorbent gloves are required to handle the workpiece. When liquid gases are used, adequate ventilation must be provided so the oxygen supply in the work area is not depleted.

When chilled fluids are used, tool life increases up to fivefold have been reported with aerospace materials. Whether this is worth the cost of this method must be determined for the specific application.

Compressed Gases as Cutting Fluids

Compressed gases can be useful as cutting fluids for certain applications. Shop air blast is common as a cutting fluid in contour sawing and is also useful in other operations. Compressed air does not have the cooling or lubricating properties of liquid fluids, but it does cool more than "dry" cutting (where air serves as a cutting fluid). In addition, compressed air does not contaminate the workpiece or the chips. This can be a problem with certain materials. It also does a good job of keeping the work area free of chips.

Other compressed gases such as carbon dioxide (CO_2) have been successfully used to improve tool life, but they suffer from the same high cost problem as chilled liquids. Bottled gas, under very high pressures, is expanded near the point of cutting by forcing it through a nozzle. The expansion of the gas cools it to temperatures below -100°F (-73.3°C). No pressure regulators are used. In the case of CO_2, dry ice may be formed momentarily on the work before it draws the heat and sublimes. The nozzles must be pointed at the cutting zone and placed as close as possible to minimize heat transfer and protect the operator. Tool life improvements up to 10 times have been reported with compressed CO_2. However, with chilled fluids, the economics involved must be carefully evaluated for the particular operation.

CHAPTER 4
SELECTING FLUIDS FOR MACHINING AND GRINDING PROCESSES

Fluids should be selected for a job based on the three primary criteria of compatibility, acceptability (see Chapters 5 and 7), and machinability. These factors are complexly interrelated which makes a formula or absolute selection rules impossible to state without numerous and confusing qualifications.

Table 4-1 plots groups of commonly used machining and grinding operations of comparable work severity against material groups of approximately equal machinability. In this table, the most important variables in the fluid selection procedure are laid out. By following the coordinates describing the job under evaluation to the point where they meet, the most efficient fluid, method of application, and surface finish can be determined.

Table 4-1, based on years of accumulated experience, is intended to serve as a reliable *guide* to fluid selection and application. For specific metal cutting operations, workpiece materials[1], type of tool used, and applications, the user must temper the generalized recommendations in Table 4-1 with his knowledge of the physics of metal cutting and the desired function of the cutting fluid. For instance, "Care should be used in selecting cutting fluids used in the machining of components highly stressed in service. For such parts, use only cutting fluids not making the part vulnerable to stress corrosion attack in the event surface films were retained on the surface. An alternative is to provide unusual care in washing the cutting fluid away after machining, with due consideration being given to the difficulties of washing complex finished assemblies"(1). See, also, Chapter 5, "Metallurgical and Chemical Compatibility."

---

[1] No water mixtures of any kind should be used when machining or grinding magnesium.

Table 4-1. Recommendations for Cutting and Grinding Fluids and Application Methods (1) (2) (3)

Material Identification Groups	Type of Cut	Turn, Bore, Face, Groove, Form, Cutoff, Box Turn, Trepan		Face Mill		End Mill	
		HSS	Carbide	HSS	Carbide	HSS	Carbide
Low-Medium Carbon Free Machining Steels[1]	* All	F 3,1-3,12,2,1-2,10	F 0,3-3,12,2,2-2,10	F 1-1,4,1,10-1,12	F 0,2-2,10,1,10-1,12	F 1-1,4,2,3-2,10	F 3,1-3,3,3,8-3,12
Low Alloy Structural Steels (Martensitic)[2]	Rough Finish	F 1-1,16,2,-2,10 3,1-3,7	F 0,1-1,21 0,3-1-3,13	F 1-1,21 1-1,21	F 0,2,1-2,10 0,3,1-3,12	F 1-1,9,2,1-2,10 3,4-3,12	F 2,1,2,2 3,1-3,12
Hot Work Die Steels (Martensitic)[3]	Rough Finish	F 1-1,21,1,28 1-1,21,1,28	F 0,1,10-1,12,1,28 0,3,-1-3,12,1,28	F 2,3-2,10,1,28 2,3-2,10,1,28	F 2,1-2,10,1,28 3,1-3,12,1,28	F 1-1,12,1,28 3,4-3,7,1,28	F 2,1-2,10,1,28 3,1-3,12,1,28
Stainless Steels (Austenitic)[4]	Rough Finish	F 1,10-1,16,1,31 3,4-3,7,2,13	F 0,2,1-2,10,1,31 0,2,1-2,10,1,31	F 2,3-2,10,2,13 3,4-3,7,2,13	F 0,2,2,2,13 3,1-3,3,2,13	F 2,3-2,10,1,31 3,4-3,7,2,13	F 0,2,2,2,13 3,1-3,3,2,13
Stainless Steels (Martensitic)[5]	Rough Finish	F 1,10-1,16,1,31 3,4-3,7,2,13	F 0,2,1-2,10,1,31 0,2,1-2,10,1,31	F 2,3-2,10,2,13 3,4-3,7,2,13	F 0,2,2,2,13 3,1-3,3,2,13	F 2,3-2,10,1,31 3,4-3,7,2,13	F 0,2,2,2,13 3,1-3,3,2,13
Precipitation Hardening Stainless Steels[6]	Rough Finish	F 1,10-1,16,1,31 3,4-3,7,2,13	F 0,2,1-2,10,1,31 0,2,1-2,10,1,31	F 2,3-2,10,2,13 3,4-3,7,2,13	F 0,2,2,2,13 3,1-3,3,2,13	F 2,3-2,10,1,31 3,4-3,7,2,13	F 0,2,2,2,13 3,1-3,3,2,13
Maraging Steels[7]	Rough Finish	F 1,13-1,21,1,29 2,3-2,10,2,12	F 0,1,13-1,21,1,29 0,2,3-2,10,1,29	F 1,13-1,21,1,29 2,3-2,10,2,12	F 0,1,13-1,21,1,29 0,2,3-2,10,1,29	F 1,13-1,21,1,29 2,3-2,10,2,12	F 0,1,13-1,21,1,29 0,2,3-2,10,1,29
Nickel Base Alloys[8]	Rough Finish	F 1,13-1,21,1,29 2,3-2,10,2,12	F 0,1,13-1,21,1,29 0,2,3-2,10,1,29	F 1,13-1,21,1,29 2,3-2,10,2,12	F 0,1,13-1,21,1,29 0,2,3-2,10,1,29	F 1,13-1,21,1,29 2,3-2,10,2,12	F 0,1,13-1,21,1,29 0,2,3-2,10,1,29
Cobalt Base Alloys[9]	Rough Finish	F 1,13-1,21,1,30 2,3-2,10,2,12	F 0,1,13-1,21,1,30 0,2,3-2,10,1,30	F 1,13-1,21,1,30 2,3-2,10,2,12	F 0,1,13-1,21,1,30 0,2,3-2,10,1,30	F 1,13-1,21,1,30 2,3-2,10,2,12	F 0,1,13-1,21,1,30 0,2,3-2,10,1,30
Cast Iron[10]	Rough Finish	M-F 2,1-2,10,1,24 3,1-3,3,2,1-2,2	M-F 0,2,3-2,10,1,24 0,3,1-3,3,2,1-2,2	M-F 2,3-2,10 3,8-3,12	M-F 0,2,3-2,10 3,8-3,12	M-F 2,3-2,10 3,8-3,12	M-F 0,2,3-2,10 3,8-3,12

SELECTING FLUIDS FOR MACHINING AND GRINDING PROCESSES 65

Table 4-1. Recommendations for Cutting and Grinding Fluids and Application Methods (1) (2) (3)—Cont'd

Material								
Magnesium and Alloys[11]	All	M-F 0,1.23,1.26,3.20	M-F 0,1.23,1.26,3.20	M-F 0,1.22,1.23,1.26	M-F 0,1.22,1.23,1.26	M-F 0,1.22,1.23,1.26	M-F 0,1.22,1.23,1.26	
Aluminum and Alloys[12]	All	M-F 3.3,2.1,1.26,3.20	M-F 3.3,2.1,1.26,3.20	M-F 3.3,2.1,1.26	M-F 3.3,2.1,1.26	M-F 3.1,3.3,2.1,1.26	M-F 3.1-3.3,2.1,1.26	
Copper and Alloys[13]	Rough	F 2.3-2.10,1.24,1.25	F 2.1,2.2,1.25,3.1	F 2.1,2.2,1.25	F 2.1,2.2,1.25	F 2.1,2.2,1.25	F 2.1,2.2,1.25	
	Finish	2.3-2.10,1.24,1.25	2.1,2.2,1.25,3.1	2.1,2.2,1.25	2.1,2.2,1.25	2.1,2.2,1.25	2.1,2.2,1.25	
Titanium and Alloys[14]	Rough	M-F 2.10,3.4,2.11	M-F 1.24,2.11,1.29	M-F 2.10,3.4,2.11	M-F 1.24,2.11,1.27	M-F 2.10,3.4,2.11	M-F 1.24,2.11,1.27	
	Finish	2.10,3.4,2.11	1.24,2.11,1.29	2.10,3.4,2.11	1.24,2.11,1.27	2.10,3.4,2.11	1.24,2.11,1.27	
Beryllium and Alloys[15]	Rough	M-F 0,2.3-2.10	M-F 0,2.2-2.2	M-F 0,2.3-2.10	M-F 0,2-2.2	M-F 0,2.2-2.10	M-F 0,2-2.2	
	Finish	0,3.4-3.7	0,3.1-3.3	0,3.4-3.7	0,3.1-3.3	0,3.4-3.7	0,3.1-3.3	
Refractories[16]	All	F 2.3-2.10,1.13-1.16	F 3.4-3.7,1.10-1.12	F 2.3-2.10,1.13-1.16	F 2.3-2.10,1.13-1.16	F 2.3-2.10,1.13-1.16	F 2.3-2.10,1.13-1.16	

The following are examples (only) of the material groups listed in Table IV-1 and are not all-inclusive.
[1]10xx Plain resulfurized carbon steel; 11xx; 13xx Manganese steel; 23xx Nickel steel; 25xx; 31xx Nickel-chrome steel.
[2]4130, 4135, 4140; 4340; Hy-Tuf (™Crucible Steel Company of America); AMS6304; 17-22 AS (14 MV); 14 CMV (Chromalloy).
[3]Vascojet 1000 (™Vanadium-Alloys Steel Company); Thermold 5 (™Universal Cyclops Specialty Steel Division, Cyclops Corporation); Patomac M (™Allegheny Ludlum Steel Corporation); Super Tricent (™The International Nickel Co., Inc.); Tool steels; Halcomb 218 (™Crucible Steel Company of America); Peerless 56 (™Crucible Steel Company of America); UHS 260; Unimach 2 (™Universal Cyclops Specialty Steel Division, Cyclops Corporation).
[4]A286; 19-9DC; N-155 Multimet Alloy (™Union Carbide Corporation, Materials Systems Division); 301 Cres; 302; 304.
[5]403 Cres; 410 Cres; 416 F. Cres; 420 Cres; 422; 430 F. Cres; 431; 440 C. Cres.
[6]PH 15-7 MD; 17-4 PH; 17-7 PH; (Armco Steel Corporation) AM 350; AM 355.
[7]GR 200 18% Ni; GR 250 18% Ni; GR 300 18% Ni; 25% Ni.
*Inconel, Inconel X (™Huntington Alloy Products Division. The International Nickel Company, Inc.); Inconel 700, 713C, 718, 901 (™Huntington Alloy Products Division. The International Nickel Company, Inc.); K Monel (™Huntington Alloy Products Division. The International Nickel Company, Inc.); Waspaloy (™Union Carbide Corporation, Materials Systems Division); René 41 (™Vacuum Melted Alloys, Metallurgical Products Department, General Electric Company); Hastelloy B, C, X (™Union Carbide Corporation, Materials Systems Division); Nimonic 80, 90 (™Mond Nickel Company, Ltd.); R-235.
[9]Haynes 25 (L-605), 21, 31 (™Union Carbide Corporation, Materials Systems Division); S-186; GE 1570.
[10]Gray; Ductile; Malleable.
[11]AZ 31, AM 240, ANM 28, AM 74S.
[12]2020, 2024, 7075, 7178.
[13]Brasses; Bronzes; Muntz metal; Nickel silver.
[14]8Al-1 Mo-IV, 6Al-4V-2Sn, 3Al-13V-11Cr.
[15]Various trade names.
[16]Molybdenum, Tungsten.
*Letters indicate methods of application (see Table IV-3).

Table 4-1. Recommendations for Cutting and Grinding Fluids and Application Methods (1) (2) (3)—Cont'd

Material Identification Groups	Type of Cut	Other Milling: Slab, Slot, Mill Saw, Hollow Mill, Thread		Drill		Gundrill	
		HSS	Carbide	HSS	Carbide	HSS	Carbide
Low-Medium Carbon Free Machining Steels[1]	All	F 3,4-3,12,1,10-1,12	F 3,8-3,12,2,1-2,2	F 3,4-3,12,1,5-1,16	F 3,4-3,12,2,3-2,10	P 2,3-2,10,1,10-1,12	P 2,3-2,10,1,10-1,12
Low Alloy Structural Steels (Martensitic)[2]	Rough Finish	F 1-1,16,2-2,10 3,1-3,7	F 2,1,2,2 3,1-3,12	1,5-1,21 3,4-3,7,2,3-2,10	F 3,4-3,12,2,3-2,10 3,4-3,12,2,3-2,10	P 1,5-1,21 1,5-1,21	P 1-1,4,1,10-1,16 1-1,4,1,10-1,16
Hot Work Die Steels (Martensitic)[3]	Rough Finish	F 1-1,21,1,28 1-1,21,1,28	F 1-1,4,1,28 3,1-3,12,1,28	F,H 1,5-1,21 3,4-3,7,2,3-2,10	F,H 3,4-3,12,1,10-1,16 3,4-3,12,1,10-1,16	P 1,5-1,21 1,5-1,21	P 1-1,4,1,10-1,16 1-1,4,1,10-1,16
Stainless Steels (Austenitic)[4]	Rough Finish	F 2,3-2,10,1,31 2,3-2,10,2,13	F 2,3-2,10,1,31 2,3-2,10,1,31	F,H 1,5-1,16 1,5-1,16	F,H 3,4-3,12,2,3-2,10 3,4-3,12,2,3-2,10	P 1,5-1,16 1,5-1,16	P 1,10-1,12 3,4-3,12
Stainless Steels (Martensitic)[5]	Rough Finish	F 2,3-2,10,1,31 2,3-2,10,2,13	F 2,3-2,10,1,31 2,3-2,10,1,31	F,H 1,5-1,16 1,5-1,16	F,H 3,4-3,12,2,3-2,10 3,4-3,12,2,3-2,10	P 1,5-1,16 1,5-1,16	P 1,10-1,12 3,4-3,12
Precipitation Hardening Stainless Steels[6]	Rough Finish	F 2,3-2,10,1,31 2,3-2,10,2,13	F 2,3-2,10,1,31 2,3-2,10,1,31	F,H 1,13-1,16 1,13-1,16	F,H 1,10-1,12 2,3-2,10	P 1,13-1,16 1,13-1,16	P 1,10-1,12 2,3-2,10
Maraging Steels[7]	Rough Finish	F 1,13-1,21,1,29 2,3-2,10,2,12	F 1,13-1,21,1,29 2,3-2,10,1,29	F,H 1,13-1,21,1,29 1,13-1,21,1,29	F,H 1,13-1,21,1,29 1,13-1,21,1,29	P 1,17-1,21,1,29 1,17-1,21,1,29	P 1,13-1,21,1,29 1,13-1,21,1,29
Nickel Base Alloys[8]	Rough Finish	F 1,13-1,21,1,29 2,3-2,10,2,12	F 1,13-1,21,1,29 2,3-2,10,1,29	F,H,I 1,13-1,21,1,29 1,13-1,21,1,29	F,H,I 1,13-1,21,1,29 1,13-1,21,1,29	P 1,17-1,21,1,29 1,17-1,21,1,29	P 1,13-1,21,1,29 1,13-1,21,1,29
Cobalt Base Alloys[9]	Rough Finish	F 1,13-1,21,1,30 2,3-2,10,2,12	F 1,13-1,21,1,30 2,3-2,10,1,30	F,H,I 1,13-1,21,1,30 1,13-1,21,1,30	F,H,I 1,13-1,21,1,30 1,13-1,21,1,30	P 1,17-1,21,1,30 1,17-1,21,1,30	P 1,13-1,21,1,30 1,13-1,21,1,30
Cast Iron[10]	Rough Finish	M-F 2,3-2,10 3,8-3,12	M-F 0,2,3-2,10 3,8-3,12	F 2,3-2,10 3,4-3,12	F 0,2,1,2,2 3,1-3,3,3,8-3,12	P 1,24 1,24	P 1,24 1,24

SELECTING FLUIDS FOR MACHINING AND GRINDING PROCESSES 67

Table 4-1. Recommendations for Cutting and Grinding Fluids and Application Methods (1) (2) (3)—Cont'd

Material		M-F	M-F	F	F	P	P
Magnesium and Alloys[11]	All	0,1,22,1.23,1.26	0,1,22,1.23,1.26	0,1,22,1.23,1.26	0,1,22,1.23,1.26	0,1,22,1.23,1.26	0,1,22,1.23,1.26
Aluminum and Alloys[12]	All	2,3-2,10,1.26	3,1-3.3,2,1,1.26	2,3-2,10,1.26	2,1-2,2,1.26	2,3-2,10,1.26	2,3-2,10,1.26
Copper and Alloys[13]	Rough	2,1-2,10,1.25	2,1-2,2,1.25	F,H 2,3-2,10,1.25	2,1-2,2,1.25	1.24	1.24
	Finish	2,1-2,10,1.25	2,1-2,2,1.25	2,3-2,10,1.25	2,1-2,2,1.25	1.24	1.24
Titanium and Alloys[14]	Rough	2,10,3,4,2.11	1,24,2,11,1.27	F,H,I 1.27	1.27	1.27	1.27
	Finish	2,10,3,4,2.11	1,24,2,11,1.27	1.27	1.27	1.27	1.27
Beryllium and Alloys[15]	Rough	0,1,5-1.9,2,3-2,10	0,1-1.14,2-2.2	2,3-2,10,1.5-1.9	2,3-2,10,1-1.4	2,3-2,10,1.5-1.9	2,3-2,10,1-1.4
	Finish	0,3,4-3.7	0,3,1-3.3	2,3-2,10,1.5-1.9	2,3-2,10,1-1.4	2,3-2,10,1.5-1.9	2,3-2,10,1-1.4
Refractories[16]	All	2,3-2,10,1,13-1.16	2,3-2,10,1,10-1.12	F,H,I,V 1.13-1.16	1.13-1.16	1.13-1.16	1.13-1.16

The following are examples (only) of the material groups listed in Table IV-1 and are not all-inclusive.

[1] 10xx Plain resulfurized carbon steel; 11xx; 13xx Manganese steel; 23xx Nickel steel; 25xx; 31xx Nickel-chrome steel.

[2] 4130, 4135, 4140; 4340; Hy-Tuf (™Crucible Steel Company of America); AMS6304; 17-22 AS (14 MV); 14 CMV (Chromalloy).

[3] Vascojet 1000 (™Vanadium-Alloys Steel Company); Thermold 5 (™Universal Cyclops Specialty Steel Division, Cyclops Corporation); Patomac M (™Allegheny Ludlum Steel Corporation); Super Tricent (™The International Nickel Co., Inc.); Tool steels; Halcomb 218 (™Crucible Steel Company of America); Peerless 56 (™Crucible Steel Company of America); UHS 260; Unimach 2 (™Universal Cyclops Specialty Steel Division, Cyclops Corporation).

[4] A286; 19-9DC; N-155 Multimet Alloy (™Union Carbide Corporation, Materials Systems Division); 301 Cres; 302; 304.

[5] 403 Cres; 410 Cres; 416 F. Cres; 420 Cres; 422; 430 F. Cres; 431; 440 C. Cres.

[6] PH 15-7 MD; 17-4 PH; 17-7 PH; (Armco Steel Corporation) AM 350; AM 355.

[7] GR 200 18% Ni; GR 250 18% Ni; GR 300 18% Ni; 25% Ni.

[8] Inconel, Inconel X (™Huntington Alloy Products Division, The International Nickel Company, Inc); Inconel 700, 713C, 718, 901 (™Huntington Alloy Products Division, The International Nickel Company, Inc.); K Monel (™Huntington Alloy Products Division, The International Nickel Company, Inc.); Waspaloy (™Union Carbide Corporation, Materials Systems Division); René 41 (™Vacuum Melted Alloys, Metallurgical Products Department, General Electric Company); Hastelloy B, C, X (™Union Carbide Corporation, Materials Systems Division); Nimonic 80, 90 (™Mond Nickel Company, Ltd.); R-235.

[9] Haynes 25 (L-605), 21, 31 (™Union Carbide Corporation, Materials Systems Division); S-186; GE 1570.

[10] Gray; Ductile; Malleable.

[11] AZ 31, AM 240, ANM 28, AM 74S.

[12] 2020, 2024, 7075, 7178.

[13] Brasses; Bronzes; Muntz metal; Nickel silver.

[14] 8Al-1 Mo-IV, 6Al-4V-25ₙ, 3Al-13V-11Cr.

[15] Various trade names.

[16] Molybdenum, Tungsten.

*Letters indicate methods of application (see Table IV-3).

68 SELECTING FLUIDS FOR MACHINING AND GRINDING PROCESSES

Table 4-1. Recommendations for Cutting and Grinding Fluids and Application Methods (1) (2) (3)—Cont'd

Material Identification Groups	Type of Cut	Spot Face, Countersink, Ream, Counterbore,		Broach		Tap, Thread, Chase		Grind: Surface, Cylindrical, Internal, Centerless
		HSS	Carbide	HSS	Carbide	HSS		
Low-Medium Carbon Free Machining Steels[1]	* All	F,H 2.1-2.2,1-1.12	F,H 0,1-1.12	F 2.3-2.10,1.5-1.16	F 2.3-2.10,1-1.12	F,H 1-1.2		F 3.1-3.3,3.8-3.12
Low Alloy Structural Steels (Martensitic)[2]	Rough Finish	F,H 2.1-2.2,1-1.21 3.1-3.3,2.1-2.2	F,H 0,1-1.12 0,3-1.3.12	F 1.5-1.16 2.1-2.2,1.5-1.16	F 1-1.4,2.3-2.10 1-1.4,2.3-2.10	F,H 1-1.16 1-1.16		F 3.1-3.3,3,8-3.12 1-3.3,3,8-3.12
Hot Work Die Steels (Martensitic)[3]	Rough Finish	F,H 2.3-2.10,1-1.21 2.3-2.10,1-1.21	F,H 1-1.16 3.1-3.12	F 1.5-1.21 1.5-1.21	F 1.5-1.16 1.5-1.16	F,H,I 1.5-1.21 1.5-1.21		F 1-1.16 3.1-3.12
Stainless Steels (Austenitic)[4]	Rough Finish	F,H 1.5-1.21 2.3-2.10	F,H 1.10-1.12 2.3-2.10	F 1.5-1.21 1.5-1.21	F 2.3-2.10,1.13-1.21 2.3-2.10,1.13-1.21	F,H,I 2.3-2.10,1.13-1.21 2.3-2.10,1.13-1.21		F 3.1-3.3,3,8-3.12 3.1-3.3,3,8-3.12
Stainless Steels (Martensitic)[5]	Rough Finish	F,H 1.5-1.21 2.3-2.10	F,H 1.10-1.12 2.3-2.10	F 1.5-1.21 1.5-1.21	F 2.3-2.10,1.13-1.21 2.3-2.10,1.13-1.21	F,H,I 1.17-1.21 1.17-1.21		F 3.1-3.3,3,8-3.12 3.1-3.3,3,8-3.12
Precipitation Hardening Stainless Steels[6]	Rough Finish	F,H 1.10-1.21 2.3-2.10	F,H 1.10-1.12 2.3-2.10	F 1.17-1.21 1.17-1.21	F 2.3-2.10,1.13-1.21 2.3-2.10,1.13-1.21	F,H,I 1.17-1.21 1.17-1.21		F 3.1-3.3,3,8-3.12 3.1-3.3,3,8-3.12
Maraging Steels[7]	Rough Finish	F,H 1.13-1.21,1.29 1.13-1.21,1.29	F,H 1.13-1.21,1.29 1.13-1.21,1.29	F 1.13-1.21,1.29 1.13-1.21,1.29	F 1.13-1.21,1.29 1.13-1.21,1.29	F,H,I 1.17-1.21,1.29 1.17-1.21,1.29		F 3.8-3.12,1.13-1.16 3.8-3.12,1.13-1.16
Nickel Base Alloys[8]	Rough Finish	F,H 1.13-1.21,1.29 1.13-1.21,1.29	F,H 1.13-1.21,1.29 1.13-1.21,1.29	F 1.13-1.21,1.29 1.13-1.21,1.29	F 1.13-1.21,1.29 1.13-1.21,1.29	F,H,I 1.17-1.21,1.29 1.17-1.21,1.29		F 3.8-3.12,1.13-1.16 3.8-3.12,1.13-1.16
Cobalt Base Alloys[9]	Rough Finish	F,H 1.13-1.21,1.30 1.13-1.21,1.30	F,H 1.13-1.21,1.30 1.13-1.21,1.30	F 1.13-1.21,1.30 1.13-1.21,1.30	F 1.13-1.21,1.30 1.13-1.21,1.30	F,H,I 1.17-1.21,1.30 1.17-1.21,1.30		F 3.8-3.12,1.13-1.16 3.8-3.12,1.13-1.16
Cast Iron[10]	Rough Finish	F,H 3.8-3.12,2.1-2.10 3.1-3.7	F,H 0,2.1,2.2 3.1-3.3,3,8-3.12	F 3.4-3.12,2.3-2.10 3.4-3.12,2.3-2.10	F 0,3.4-3.12,2.3-2.10 0,3.4-3.12,2.3-2.10	F,H 0,3.4-3.12,2.3-2.10 0,3.4-3.12,2.3-2.10		F 3.8-3.12,2.1-2.10 3.1-3.7

Table 4-1. Recommendations for Cutting and Grinding Fluids and Application Methods (1) (2) (3)—Cont'd

Material		F,H	F,H	F	F,H	F
Magnesium and Alloys[11]	All	0,1,22,1,23,1,26	0,1,22,1,23,1,26	0,1,22,1,23,1,26	0,1,22,1,23,1,26	0,1,22,1,26
Aluminum and Alloys[12]	All	2.1-2,2,1.26	2.1-2,2,1.26	2.3-2,10,1.26	2.3-2,10,1.26	2.3-2,10,1.26
Copper and Alloys[13]	Rough	2,3-2,10,1.24,1.25	2.1-2,2,1.24	2.3-2,10,1.24	2.3-2,10,1.24	3,4-3,7,1.25
	Finish	2,3-2,10,1.24,1.25	2.1-2,2,1.24	2.3-2,10,1.24	2.3-2,10,1.24	3,4-3,7,1.25
Titanium and Alloys[14]	Rough	2.11	1,27	1,27	F,P,H,I 1,34,1.27	3,14,3,15,2,11
	Finish	2.11	1,27	1,27	1,34,1.27	3,14,3,15,2,11
Beryllium and Alloys[15]	Rough	0,2,1-2,10	0,2,1-2,10	2,3-2,10,1.5-1.9	F,P,H,I 2,3-2,10,1.5-1.9	0,2,1-2,2
	Finish	0,2,1-2,10	0,2,1-2,10	2,3-2,10,.5-1.9	2,3-2,10,1.5-1.9	3,1-3,3
Refractories[16]	All	2,3-2,10,1,13-1,16	2,3-2,10,1,13-1,16	1,13-1,16	F,H,P,I 1,13-1,16	1,10-1,16,3,15

The following are examples (only) of the material groups listed in Table IV-1 and are not all-inclusive.

[1]0xx: Plain resulfurized carbon steel; 11xx: 13xx Manganese steel; 23xx Nickel steel; 25xx: 31xx Nickel-chrome steel.

[2]4136, 4135, 4140; 4340; Hy-Tuf (™Crucible Steel Company of America); AMS6304; 17-22 AS (14 MV); 14 CMV (Chromalloy).

[3]Vascojet 1000 (™Vanadium-Alloys Steel Company); Thermold 5 (™Universal Cyclops Specialty Steel Division, Cyclops Corporation); Patomac M (™Allegheny Ludlum Steel Corporation); Super Tricent (™The International Nickel Co., Inc.); Tool steels; Halcomb 218 (™Crucible Steel Company of America); Peerless 56 (™Crucible Steel Company of America); UHS 260; Unimach 2 (™Universal Cyclops Specialty Steel Division, Cyclops Corporation).

[4]A286; 19-9DC; N-155 Multimet Alloy (™Union Carbide Corporation, Materials Systems Division); 301 Cres; 302; 304.

[5]403 Cres; 410 Cres; 416 F. Cres; 420 Cres; 422; 430 F. Cres; 431; 440 C. Cres.

[6]PH 15-7 MD; 17-4 PH; 17-7 PH; (Armco Steel Corporation) AM 350; AM 355.

[7]GR 200 18% Ni; GR 250 18% Ni; GR 300 18% Ni; 25% Ni.

[8]Inconel, Inconel X (™Huntington Alloy Products Division, The International Nickel Company, Inc.); Inconel 700, 713C, 718, 901 (™Huntington Alloy Products Division, The International Nickel Company, Inc.); K Monel (™Huntington Alloy Products Division, The International Nickel Company, Inc.); Waspaloy (™Union Carbide Corporation, Materials Systems Division); René 41 (™Vacuum Melted Alloys, Metallurgical Products Department, General Electric Company); Hastelloy B, C, X (™Union Carbide Corporation, Materials Systems Division); Nimonic 80, 90 (™Mond Nickel Company, Ltd.); R-235.

[9]Haynes 25 (L-605), 21, 31 (™Union Carbide Corporation, Materials Systems Division); S-186; GE 1570.

[10]Gray; Ductile; Malleable.

[11]AZ 31, AM 240, ANM 28, AM 74S.

[12]2020, 2024, 7075, 7178.

[13]Brasses; Bronzes; Muntz metal; Nickel silver.

[14]8Al-1 Mo-1V, 6Al-4V-2S$_n$, 3Al-13V-11Cr.

[15]Various trade names.

[16]Molybdenum; Tungsten.

*Letters indicate methods of application (see Table IV-3).

70 SELECTING FLUIDS FOR MACHINING AND GRINDING PROCESSES

Table 4-1. Recommendations for Cutting and Grinding Fluids and Application Methods (1) (2) (3)—Cont'd

Material Identification Groups	Type of Cut	Grind: Threads, Gear, Form	Gear: Hob, Cut, Shape, Shave	Saw: Power, Hack, Cutoff HSS	Saw: Power, Hack, Cutoff Carbide	Saw: Abrasive	Hone
Low-Medium Carbon Free Machining Steels[1]	* All	F 3.4-3.12,1.10-1.12	F 2.3-2.10,1.10-1.12	F 2.3-2.10,1.10-1.12	F 2.3-2.10,1-1.4	F 3.1-3.12	F 1.10-1.12,1.33
Low Alloy Structural Steels (Martensitic)[2]	Rough Finish	F 1-1.21 3.1-3.12	F 1-1.21 2.3-2.10	F 1-1.21 2.3-2.10	F 1-1.12 2.3-2.10	F 3.1-3.12 3.1-3.12	F 1.10-1.12,1.33 1.10-1.12,1.33
Hot Work Die Steels (Martensitic)[3]	Rough Finish	F 1-1.16 1-1.16	F 1.5-1.21 1.5-1.21	F 1-1.16 2.3-2.10	F 1-1.12 3.1-3.12	F 3.1-3.12 3.1-3.12	F 1.10-1.12,1.33 1.10-1.12,1.33
Stainless Steels (Austenitic)[4]	Rough Finish	F 1.10-1.16 1.10-1.16	F 1.13-1.16 1-1.4	F 1.13-1.16 2.3-2.10	F 1.10-1.12 2.3-2.10	F 1-1.12 3.1-3.12	F 1.10-1.12,1.33 1.10-1.12,1.33
Stainless Steels (Martensitic)[5]	Rough Finish	F 1.10-1.16 1.10-1.16	F 1.13-1.16 1-1.4	F 1.13-1.16 2.3-2.10	F 1.10-1.12 2.3-2.10	F 1-1.12 3.1-3.12	F 1.10-1.12,1.33 1.10-1.12,1.33
Precipitation Hardening Stainless Steels[6]	Rough Finish	F 1.10-1.16 1.10-1.16	F 1.13-1.16 1.13-1.16	F 1.13-1.16 1.13-1.16	F 1.10-1.12 1.10-1.12	F 1.13-1.16 3.4-3.12	F 1.10-1.12,1.33 1.10-1.12,1.33
Maraging Steels[7]	Rough Finish	F 1.13-1.21,1.29 1.13-1.21,1.29	F 1.13-1.21,1.29 1.13-1.21,1.29	F 1.13-1.21,1.29 1.13-1.21,1.29	F 1.13-1.21,1.29 1.13-1.21,1.29	F 3.8-3.12,1.13-1.16 3.8-3.12,1.13-1.16	F 1.10-1.12,1.33 1.10-1.12,1.33
Nickel Base Alloys[8]	Rough Finish	F 1.13-1.21,1.29 1.13-1.21,1.29	F 1.13-1.21,1.29 1.13-1.21,1.29	F 1.13-1.21,1.29 1.13-1.21,1.29	F 1.13-1.21,1.29 1.13-1.21,1.29	F 3.8-3.12,1.13-1.16 3.8-3.12,1.13-1.16	F 1.10-1.12,1.33 1.10-1.12,1.33
Cobalt Base Alloys[9]	Rough Finish	F 1.13-1.21,1.30 1.13-1.21,1.30	F 1.13-1.21,1.30 1.13-1.21,1.30	F 1.13-1.21,1.30 1.13-1.21,1.30	F 1.13-1.21,1.30 1.13-1.21,1.30	F 3.8-3.12,1.13-1.16 3.8-3.12,1.13-1.16	F 1.10-1.12,1.33 1.10-1.12,1.33
Cast Iron[10]	Rough Finish	F 3.1-3.3,2.1,2.2 3.8-3.12	F 2.3-2.10,1.24 3.4-3.12	F 3.4-3.12,2.3-2.10 3.4-3.12,2.3-2.10	F 3.8-3.12,2.1-2.2 3.1-3.3,3.8-3.12	F 3.8-3.12,2.3-2.10 2.3-2.10,3.4-3.12	F 1.24,1.33 1.24,1.33

Table 4-1. Recommendations for Cutting and Grinding Fluids and Application Methods (1) (2) (3)—Cont'd

Material	Sub						
Magnesium and Alloys[11]	All	F 0,1,22,1,26	F 0,1,22,1,26	F 0,1,22,1,26	F 0,1,22,1,26	F 0,1,22,1,26	F 1.33
Aluminum and Alloys[12]	All	F 2,3-2,10,1,26	F 2,3-2,10,1,26	F 2,3-2,10,1,26	F 3,1-3,3,2,1,1,26	F 2,3-2,10,1,26	F 1.33
Copper and Alloys[13]	Rough Finish	F 2,3-2,10,1,25 2,3-2,10,1,25	F 2,3-2,10,1,25 2,3-2,10,1,25	F 1,24 1,24	F 2,1,2,2,1,25 2,1,2,2,1,25	F 2,3-2,10,1,25 2,3-2,10,1,25	F 1.33 1.33
Titanium and Alloys[14]	Rough Finish	F 3,14,3,15,1,27 3,14,3,15,1,27	F 1,27 1,27	F 3,4,3,13,2,11 3,4,3,13,2,11	F 3,4,3,13,1,27 3,4,3,13,1,27	F 3,14,3,15,2,11 3,14,3,15,2,11	F 1,33,1,27 1,33,1,27
Beryllium and Alloys[15]	Rough Finish	F 2,1-2,2,1-1,4 2,1-2,2,1-1,4	F 0,2,1,2,2 0,2,1,2,2	F 2,3-2,10,1,5-1,9 2,3-2,10,1,5-1,9	F 2,1,2,2,1-1,4 2,1,2,2,1-1,4	F 0,2,1-2,2 6,3,1-3,3	F 1.33 1.33
Refractories[16]	All	F 1,10-1,16,3,15	F 1,13-1,16	F 1,13-1,16	F 1,13-1,16	F 1,10-1,12,3,15	F 1,10-1,12,3,15

The following are examples (only) of the material groups listed in Table IV-1 and are not all-inclusive.

[1] 10xx Plain resulfurized carbon steel; 11xx; 13xx Manganese steel; 23xx Nickel steel; 25xx; 31xx Nickel-chrome steel.

[2] 4130, 4135, 4140; 4340; Hy-Tuf (™Crucible Steel Company of America); AMS6304; 17-22 AS (14 MV); 14 CMV (Chromalloy).

[3] Vascojet 1000 (™Vanadium-Alloys Steel Company); Thermold 5 (™Universal Cyclops Specialty Steel Division, Cyclops Corporation); Patomac M (™Allegheny Ludlum Steel Corporation); Super Tricent (™The International Nickel Co., Inc.); Tool steels; Halcomb 218 (™Crucible Steel Company of America); Peerless 56 (™Crucible Steel Company of America); UHS 260; Unimach 2 (™Universal Cyclops Specialty Steel Division, Cyclops Corporation).

[4] A286; 19-9DC; N-155 Multimet Alloy (™Union Carbide Corporation, Materials Systems Division); 301 Cres; 302; 304.

[5] 403 Cres; 410 Cres; 416 F. Cres; 420 Cres; 422; 430 F Cres; 431; 440 C. Cres.

[6] PH 15-7 MD; 17-4 PH; 17-7 PH; (Armco Steel Corporation) AM 350; AM 355.

[7] GR 200 18% Ni; GR 250 18% Ni; GR 300 18% Ni; 25% Ni.

[8] Inconel, Inconel X (™Huntington Alloy Products Division, The International Nickel Company, Inc.); Inconel 700, 713C, 718, 901 (™Huntington Alloy Products Division, The International Nickel Company, Inc.); K Monel (™Huntington Alloy Products Division, The International Nickel Company, Inc.); Waspaloy (™Union Carbide Corporation, Materials Systems Division); René 41 (™Vacuum Melted Alloys, Metallurgical Products Department, General Electric Company); Hastelloy B, C, X (™Union Carbide Corporation, Materials Systems Division); Nimonic 80, 90 (™Mond Nickel Company, Ltd.); R-235.

[9] Haynes 25 (L-605), 21, 31 (™Union Carbide Corporation, Materials Systems Division); S-186; GE 1570.

[10] Gray; Ductile; Malleable.

[11] AZ 31, AM 240, ANM 28, AM 74S.

[12] 2020, 2024, 7075, 7178.

[13] Brasses; Bronzes; Muntz metal; Nickel silver.

[14] 8Al-1 Mo-IV, 6Al-4V-25$_n$, 3Al-13V-11Cr.

[15] Various trade names.

[16] Molybdenum, Tungsten.

* Letters indicate methods of application (see Table IV-3).

In addition, the *form* of the cutting tool is important (4) in helping to determine the primary property required of the fluid, i.e., lubricity or cooling or both. Since chips tend to flow perpendicularly to the cutting edge, the form of the tool design will determine whether the chip "piles up" upon itself or spreads out as shown in Figure 4-1.

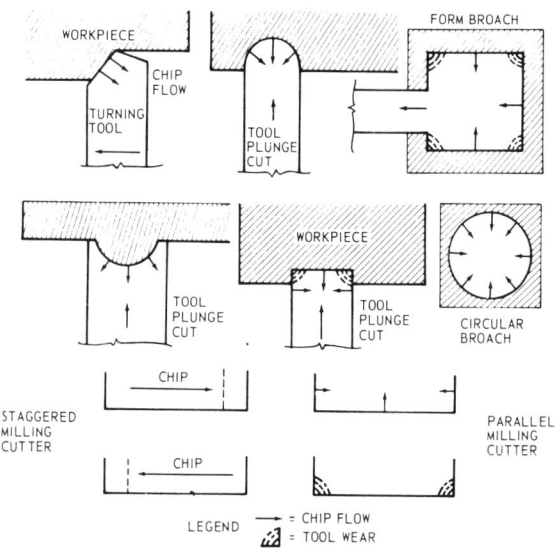

Figure 4-1. *Effect of tool form on chip flow and tool wear.*

All metals "give" and then "spring back" to a greater or lesser extent when machined. This means that on grooving cuts, cut-off, or recessing operations, good lubricity from the fluid is needed to prevent excessive tool wear at the corners. At the tool corners, chips "pile up," exerting more pressure and heat than for the same feed in a simple turning cut. These corners have very little "bulk" or strength, so they break down rather readily. Similarly, forms used in internal broaching will undergo more strain than teeth in surface broaching.

A plunge cutting tool requires less lubricity from the cutting oil if the chips spread out. Compare this with a nose tool which will "crimp" the chip on a plunge, creating less corner wear on the cutter than with a parallel tooth milling cutter where the chips "pile up" at the corners.

Tools should be ground, wherever possible, with the grinding marks parallel with the cutting edge on the tool face and flank (5,6). If the chip flows perpendicularly to the grinding marks as in Figure 4-2, the peaks of the grind marks serve as multiple, minute cutters which facilitate the overall cutting process. In addition, the valleys between the peaks hold the cutting oil better. If the tools are ground with the grinding marks perpendicular to the cutting edge, the peaks can weld more easily to the chip, increasing the load on the tool and reducing tool life.

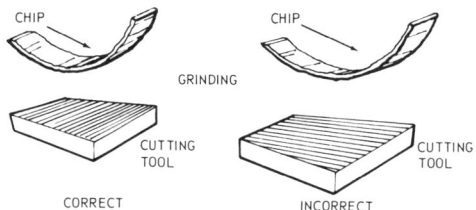

Figure 4-2. *How to grind tools for better cutting effect and cutting oil use.*

Shaw (6) indicates it is possible for active chlorine and sulfur cutting oils to react with the cobalt binder in carbide tools causing "leaching" of the binder and weakening the cutting edge. Care should be exercised in selecting an oil which will not adversely affect such tools.

Ceramic tools, because of their low conductivity and because they are run at high speeds (2,000 to 3,000 sfm) dissipate most of the heat created into the chips. Usually, ceramic tools are used without cutting fluid.

The cutting fluid types cited in Table 4-1 are based on the following breakdown of the three basic types of cutting fluids (2): (A) cutting oils, (B) emulsifiable oils (water miscible), and (C) chemical fluids.

A. Cutting Oils
1. Inactive
    a) Straight mineral oils
    b) Fatty oils
    c) Fatty oil-mineral blends
    d) Sulfurized fatty-mineral oil blends
2. Active (extreme pressure)
    a) Sulfurized mineral oils
    b) Sulfo-chlorinated mineral oils
    c) Sulfo- or sulfo-chlorinated fatty oil blends

B. Emulsified Oils (water miscible)
   a) Emulsifiable mineral oils
   b) Super fatted emulsifiable oils
   c) Extreme pressure (EP) emulsifiable oils

C. Chemical Fluids (synthetic)
   a) True solution type
   b) Wetting agent types
   c) Wetting agent type with EP lubricant

The codes used for cutting fluid types are given in Table 4-2 (3), and the codes used for application methods are given in Table 4-3 (3).

To use Table 4-1, Recommendations for Cutting and Grinding Fluids and Application Methods:

74 SELECTING FLUIDS FOR MACHINING AND GRINDING PROCESSES

Table 4-2. Cutting and Grinding Fluid Codes (3)

Code	Fluid Type	Code	Fluid Type
0	Dry	1.21	Highly chlorinated mineral lard oil, heavy duty
1.	Cutting oils (A and B types)		
1.1	Sulfurized oil*, light duty	1.22	Straight mineral oil
1.2	Sulfurized mineral-lard oil, light duty	1.23	Straight oil
1.3	Sulfurized mineral oil*, light duty	1.24	Mineral lard oil, medium heavy duty
1.4	Sulfurized lard oil with chlorine, light duty	1.25	Mineral lard oil, light duty
1.5	Sulfurized oil, medium/heavy duty	1.26	Oil specialty recommended for aluminum and magnesium alloys
1.6	Sulfurized mineral-lard oil, medium heavy duty	1.27	Oil specialty recommended for titanium alloys
1.7	Sulfurized fat compounded oil, medium/heavy duty	1.28	Oil specialty recommended for high-temperature alloys
1.8	Sulfurized mineral oil, medium/heavy duty	1.29	Oil specialty recommended for nickel base alloys
1.9	Sulfurized lard oil, medium/heavy duty	1.30	Oil specialty recommended for cobalt base alloys
1.10	Sulfo-chlorinated mineral-lard oil, light duty	1.31	Oil specialty recommended for stainless steel alloys
1.11	Sulfo-chlorinated mineral oil, light duty	1.32	Chlorinated mineral lard oil
1.12	Sulfo-chlorinated lard oil, light duty	1.33	Honing oil
		1.34	Tapping oil
1.13	Sulfo-chlorinated mineral lard oil, medium duty	2.	Emulsifiable oils (water miscible) all
1.14	Sulfo-chlorinated mineral oil, medium duty	2.1	Water miscible oil, light duty
		2.2	Water miscible oil, medium duty
1.15	Sulfo-chlorinated lard oil, medium duty	2.3	Water miscible oil, heavy duty
1.16	Sulfo-chlorinated oil, medium duty	2.4	Sulfo-chlorinated water miscible oil, heavy duty
1.17	Sulfo-chlorinated mineral lard oil, heavy duty	2.5	Chlorinated water miscible oil, heavy duty
1.18	Sulfo-chlorinated lard oil, heavy duty	2.6	Sulfo-chlorinated water miscible compound, heavy duty
1.19	Sulfo-chlorinated oil, heavy duty	2.7	Water miscible compound, active sulfur, heavy duty
1.20	Sulfo-chlorinated mineral oil, heavy duty	2.8	Water miscible mineral oil

SELECTING FLUIDS FOR MACHINING AND GRINDING PROCESSES

Table 4-2. Cutting and Grinding Fluid Codes (3) - Cont'd

Code	Fluid Type	Code	Fluid Type
2.9	Fatty water miscible oil	3.8	Chemical coolant
2.10	Extreme pressure water miscible oil, heavy duty	3.9	Chemical solution
		3.10	Chemical emulsion, oil base
2.11	Water miscible oil specialty recommended for titanium alloys	3.11	Chemical and oil solution, heavy duty
2.12	Water miscible oil specialty recommended for high nickel cobalt alloys	3.12	Chemical and organic compound solution
2.13	Water miscible oil specialty recommended for stainless steel alloys	3.13	Chemical with extreme pressure and wetting agent water miscible
3.	Chemical fluids	3.14	Amine nitrite
3.1	Chemical emulsion, light duty	3.15	5% sodium nitrite solution
3.2	Water base chemical, light duty	3.16	Mineral oil base chemical
3.3	Water miscible petrochemical, light duty	3.17	Chemical water mix compound
		3.18	Chemical emulsion, heavy duty
3.4	Chemical emulsion, heavy duty	3.19	Chemical emulsion, medium duty
3.5	Sulfurized water based chemical, heavy duty	3.20	Chemical fluid specialty for non-ferrous alloys
3.6	Chlorinated water based chemical, heavy duty	3.21	Chemical fluid specialty for ferrous alloys
3.7	Water miscible, heavy duty		

* In these tables, the sulfurized oils and sulfurized mineral oils differ as to the sources of crudes. These will be naphthenic or paraffinic and their respective chemical reactivity will vary accordingly.

Table 4-3. Application Methods (3)

Code	Application
F	**Flood** (minimum of 3 gal./min./nozzle). Low pressure/high volume application to cool tool and workpiece and supply fluid to cutting zone. Fluid should be directed into the clearance angles of the cut, and completely envelop tool and workpiece (2). Nozzles on lathe type tools should be at least ¾ of the cutting width, with the same ratio for milling operations through the use of fan nozzles. On face milling, all teeth should be immersed continuously, using ring distributor.
M	**Mist** (water miscible oils or chemical fluids). Uses air or aerosol-powered aspirating equipment to disperse fluid as very fine droplets in the carrier, which is then directed at the cutting area. Provides low volume and high velocity primarily used where flood coolant application can not be utilized. Visibility of cut is increased with reduced cooling of tool and workpiece. Added ventilation is recommended as well as automatic shut-off.
P	**High Pressure** (50-2000 psi.). Pressure systems are used normally for internal application of fluid through drills, gundrills, end mills, and grinding wheels.
V	**High Velocity Jet.** Specialized application for increased penetration of fluid into cutting area. Creates considerable mist and smoke.
H	**Hand.** Manual application of paste, solid, or liquid by brush, dipping, or oil can.
I	**Immersion.** Gravity or low pressure application by submerging workpiece in tank or receptacle.

1. Select the appropriate workpiece material group and type of cut desired, listed in the first two left-hand columns of Table 4-1.

2. Follow the material group horizontally to the machining operation desired.
3. Read the fluid and application recommendations. Refer to fluid type and application identification codes in Tables 4-2 and 4-3.

Multiple recommendations are given to allow users greater ease in matching recommendations with their existing stock and suppliers' identifications, and to cover the variations within the material and machining operation groups. The preferred recommendation is given first, followed by other recommendations in descending order of preference. Where a range of fluids is shown, all fluids within that range are recommended for the workpiece material and type of cut indicated.

The groups of machining operations included in Table 4-1 are briefly described in the following sections (7). Typical machining operations and/or the cutting and machine tools which perform these operations are illustrated in the following examples.

TURNING

A turning operation uses single point tools to generate or cut a surface of revolution--cylindrical, tapered or contoured--on a lathe, turret lathe, automatic bar machine, etc. The tool feeds into the rotating workpiece parallel to the axis of revolution of the surface being cut as in Figure 4-3.

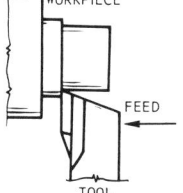

Figure 4-3. *Relationship of workpiece, tool, and feed in a turning operation.*

BORING

Boring is the operation used to enlarge a hole to an exact size with a single point tool as illustrated in Figure 4-4.

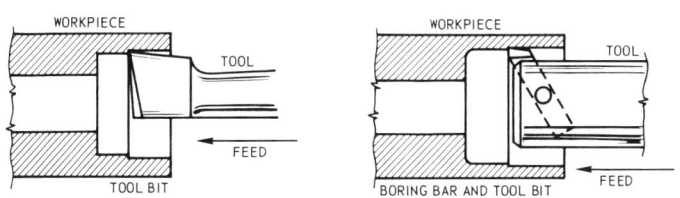

Figure 4-4. *Relationship of tools, workpieces, and feeds in a boring operation.*

FACING

Facing is the term used to describe cutting of a flat surface as the workpiece revolves in a lathe in order to square the end with the sides, or reduce the stock to a desired length. The single point tool feeds along a line perpendicular to the spindle axis as depicted in Figure 4-5.

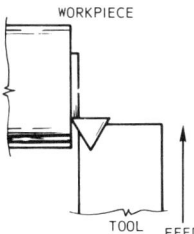

Figure 4-5. *Relationship of tool, workpiece, and feed in facing operation.*

GROOVING

Grooving is the process of machining a recess in a revolving workpiece to a specific depth, width, and shape. The single point tool, with proper groove configuration, feeds along a line perpendicular to the spindle axis as seen in Figure 4-6.

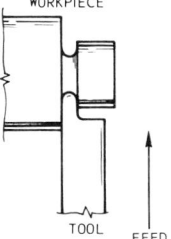

Figure 4-6. *Relationship of tool, workpiece, and feed in grooving operation.*

FORMING

A forming operation cuts curved or irregular shapes in the workpiece as it revolves in a lathe as shown in Figure 4-7.

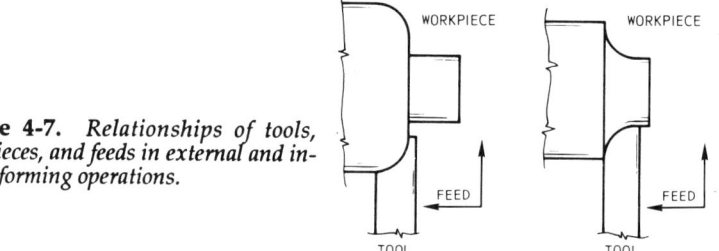

Figure 4-7. *Relationships of tools, workpieces, and feeds in external and internal forming operations.*

CUTOFF

The cutoff or parting operation is similar to the facing operation, except that the cut is made completely through the workpiece, severing one end. A cutoff tool is used similar to that shown in Figure 4-8.

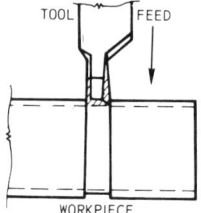

Figure 4-8. *Relationships of tools, workpieces, and feeds in cutoff operation.*

BOX TURN

The box turn turret lathe operation is similar to the turning operation except the tool(s), cutter holder, and steady rest comprise one unit as seen in Figure 4-9. The steady rest travels with the tool, supports the work, and holds the work against the tool.

 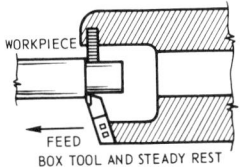

Figure 4-9. *Relationships of box tool and steady rest to workpiece and feed in box turn turret lathe operation.*

TREPANNING

Figure 4-10 shows the trepan operation using a single point tool to produce a hole by machining a circumferential groove parallel to the axis of rotation, cutting a cylindrical path and leaving a solid core. This core passes through the hollow cylindrical cutting head as the tool feeds into the metal. Trepanning is unlike the drilling operation, which reduces the metal to chips.

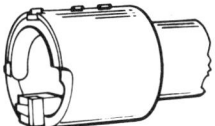

Figure 4-10. *Trepanning tool.*

MILLING

Milling is the removal of metal by a rotating cutter having one or more cutting elements called flutes or teeth. Together with the workpiece, they are held in a milling machine. The cutter rotates while the work feeds under the cutter in order to remove the metal.

Face Milling

In face milling, metal is removed by the cutting edges located on the corners and periphery of the cutter. Face mills are designed to mill flat surfaces normal to their axis of rotation as shown in Figures 4-11 and 4-12.

80 SELECTING FLUIDS FOR MACHINING AND GRINDING PROCESSES

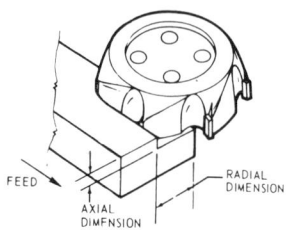

Figure 4-11. *Common face milling operation.*

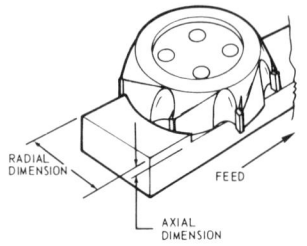

Figure 4-12. *Shallow shoulder milling operation.*

End Milling

End mills are cylindrical in shape and are provided with a shank for mounting and driving. They have straight or helical teeth on the circumferential surfaces as shown in the various operations illustrated in Figures 4-13 through 4-20.

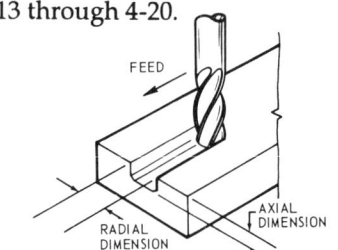

Figure 4-13. *Open end mill slotting (vertical machine).*

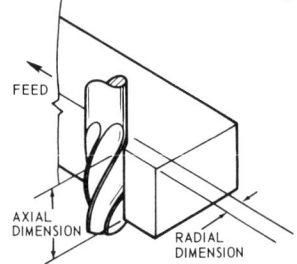

Figure 4-14. *Slab end milling (vertical machine).*

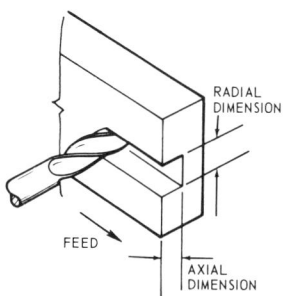

Figure 4-15. *Open end mill slotting (horizontal machine).*

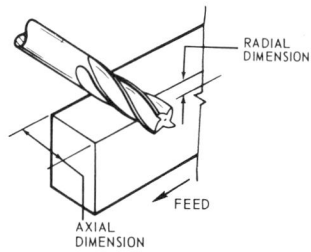

Figure 4-16. *Slab end milling (horizontal machine).*

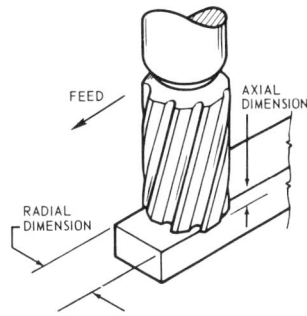

Figure 4-17. *End mill facing.*

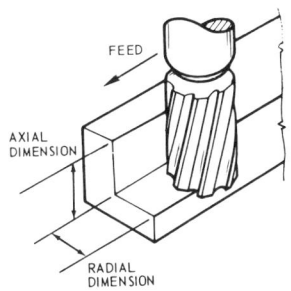

Figure 4-18. *End mill shoulder cutting.*

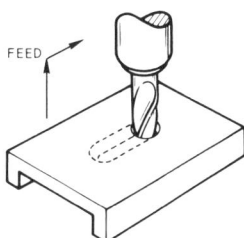

Figure 4-19. *End mill boring or drilling to make a blind spot.*

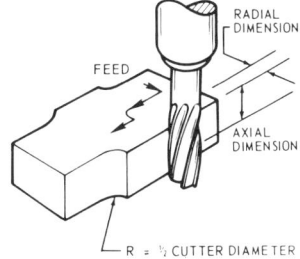

Figure 4-20. *End mill profile slab cutting.*

Slab Milling (Plain Milling)

The slab mill is a cylinder with teeth cut around the periphery only. It produces a flat surface parallel to its own axis. This feature is illustrated in figures 4-21 and 4-22.

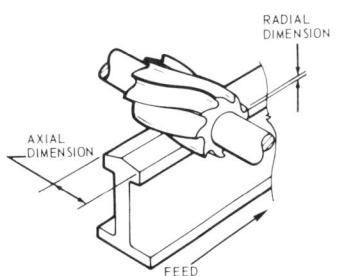

Figure 4-21. *Slab mill used for facing fragile parts.*

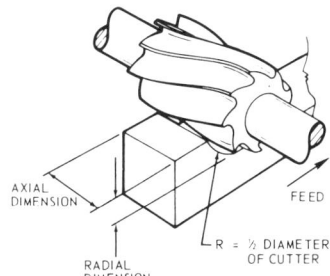

Figure 4-22. *Slab mill used for machining cuts requiring a larger radius than could be achieved by a radius ground on any type cutter.*

Side Milling

A side mill has a relatively narrow face with teeth on one or both sides, as well as on the periphery as seen in Figure 4-23. It is used for machining open slots, side milling shoulders, and straddle milling.

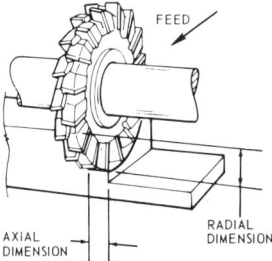

Figure 4-23. *Side milling.*

Slot Milling

A slot mill is a combination side milling cutter and end mill because it has teeth on each side, as well as on the periphery. Examples are provided in Figures 4-24 and 4-25.

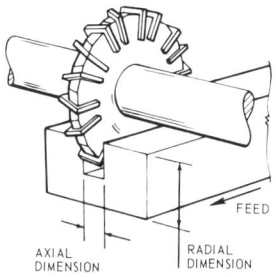

Figure 4-24. *Slotting.*

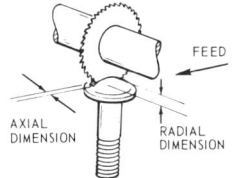

Figure 4-25. *Small slotting.*

Formed Milling Cutters

Formed cutters incorporate cam-relieved teeth which are sharpened by grinding the tooth faces and thereby maintain the original form throughout the life of the cutter. Formed cutters produce irregular or circular surfaces.

Types of formed milling cutters include convex, concave, corner-rounding, gear tooth cutters, multiple thread mills, and hobs. Typical cuts are shown in Figures 4-26 and 4-27.

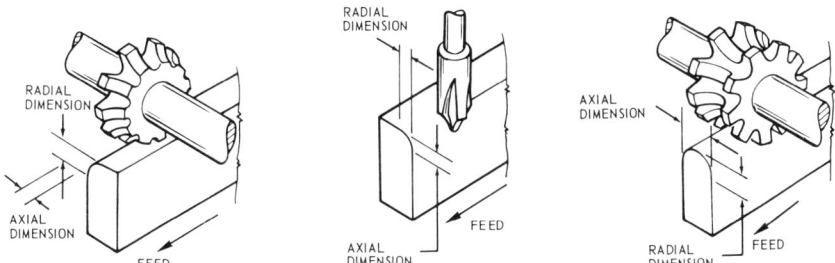

Figure 4-26. *Typical convex form cuts made by corner rounding, corner rounding end mills, and concave cutters.*

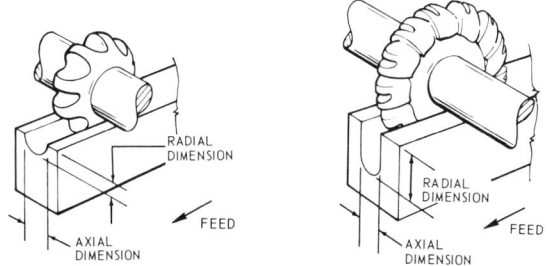

Figure 4-27. *Typical concave form cuts made by convex and full radius cutters.*

Milling Saw

A milling saw is essentially a thin plain milling cutter (teeth cut around the periphery only) that produces a flat surface parallel to its own axis as in Figure 4-28. The sides are dished to the cutting edges to provide tooth relief. It is used for cutoff work or milling very narrow slots.

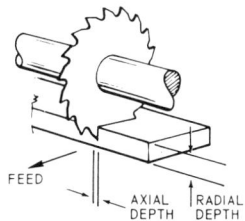

Figure 4-28. *Mill sawing.*

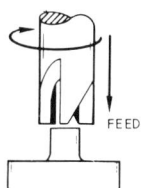

Figure 4-29. *Hollow mill.*

84 SELECTING FLUIDS FOR MACHINING AND GRINDING PROCESSES

Hollow Milling

A hollow mill is a cutter of tubular construction, as depicted in Figure 4-29, having teeth on one end and internal clearance. It is used for sizing cylindrical stock or machining straight ends of work.

Thread Milling

Figure 4-30 shows a thread milling cutter. It is a rotary cutting tool having teeth which intermittently engage the workpiece to cut internal or external threads of a specific form.

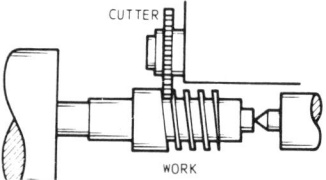

Figure 4-30. *Thread milling cutter.*

DRILLING

Drills, shown in Figure 4-31, are end cutting tools having one or more cutting edges. They incorporate helical or straight flutes for the passage of cutting fluid and chips when producing a hole.

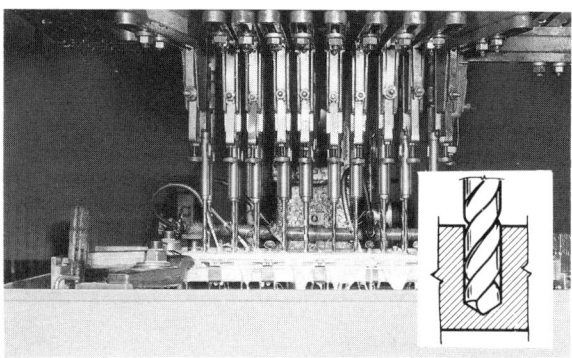

Figure 4-31. *Multiple spindle drill press. (Courtesy, Lockheed-California Company)*

GUNDRILLING

Gundrills are used to drill holes five or more times the drill diameter in depth. A gundrill is a single lip straight flute drill, illustrated in Figure 4-32, having a tubular stem through which cutting fluid is forced to flush the chips out of the flute. It is normally held stationary while the work revolves. The gundrill provides exceptional accuracy, straightness, and a quality of surface finish not otherwise obtainable.

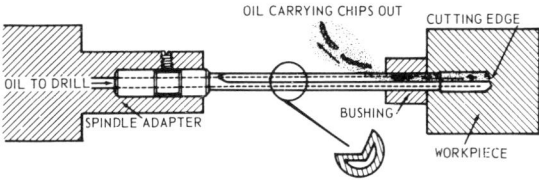

Figure 4-32. *A gun drill is essentially a single-point end cutter similar to a boring cutter.*

COUNTERBORING

The counterboring operation enlarges a previously formed hole for part of its depth, usually to produce a shoulder at the bottom of the enlargement. A guide or pilot, like the one seen in Figure 4-33, assures concentricity of the original hole with the enlarged portion. When the counterbored portion is shallow, the operation is called spot facing.

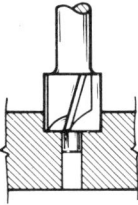

Figure 4-33. *Counter bore and pilot.*

SPOT FACING

The spot facing operation cuts a circular spot below a surface in a plane perpendicular to the axis of a hole through the spot. Spot facing and counterboring tools are end cutting tools.

COUNTERSINKING

Figure 4-34 shows the countersinking operation, using a tool with two beveled cutting elements and two flutes to countersink holes for fasteners. The cutting action is similar to the drill.

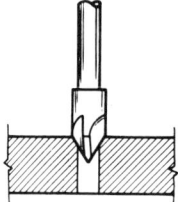

Figure 4-34. *Countersinking tool.*

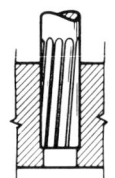

Figure 4-35. *Reamer.*

REAMING

Reamers are rotary cutting tools that enlarge and finish a previously formed hole to accurate dimensions. A reamer is an end cutting tool. It usually has more than one cutting element and flutes along the longitudinal axis. These flutes form teeth, as well as grooves, for the passage of cutting fluid and chip removal as seen in Figure 4-35. It is designed to remove a specific amount of stock (i.e., a cutting action, not a burnishing action) in order to make the hole straight, smoothly finished, and of precise dimension.

BROACHING

Broaching is a metal cutting operation combining roughing and finishing. It removes stock to precision limits with good finish quality faster than any other known metal cutting process. One of the primary reasons for this is the difference in the principle between the broaching tool and other metal cutting tools. This difference is illustrated in Figures 4-36 and 4-37. Although broaching combines roughing and finishing, no individual broach tooth handles both functions. Each successive tooth removes only a predetermined amount of stock, and is in cutting contact only a short time.

TAPPING (INTERNAL THREAD CUTTING)

A tap is a cylindrical thread cutting tool with one or more cutting elements having threads of a desired form on the periphery as shown in Figure 4-38. By a combination of rotary and axial motion, the leading edge cuts an internal thread while the tap derives its principal support from the thread it produces.

SINGLE POINT THREAD CUTTING

Single point threading is accomplished using a single point tool on a lathe or lathe-type machine as seen in Figure 4-39. Threading is produced by a combination of rotary motion of the workpiece and longitudinal motion of the carriage.

CHASING (EXTERNAL THREAD CUTTING)

A threading die (chaser) is a tool used to cut external threads. It consists of blades or circular forms made in sets and positioned with holders in a die head as shown in Figure 4-40. It cuts simultaneously on more than one point of the outside surface. Die heads are often spring-loaded and adjustable to release at full depth.

88 SELECTING FLUIDS FOR MACHINING AND GRINDING PROCESSES

Figure 4-36. *Broaching machine with broach in place ready to pull through. (Courtesy, Lockheed-California Company)*

Figure 4-37. *Horizontal pull-type broaching machine showing insertion of broach prior to cutting internal hole. (Courtesy, Lockheed-California Company).*

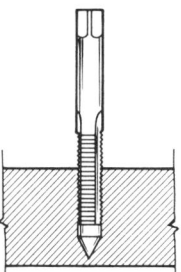

Figure 4-38. *Tap used as a cylindrical thread cutting tool.*

Figure 4-39. *Single point tool thread-cutting on conventional engine lathe. (Courtesy, Lockheed-California Company)*

GRINDING

The grinding process is a method of machining by abrasive wheels and other shapes, or by abrasive belts. Grains of extremely hard materials, ranging from microns to approximately 1/8 inch (3.175 mm) in size, serve as the cutting edges. Millions of these grains, cemented together or adhered to cloth, pass across the surface of the workpiece at high speeds to remove stock in substantial amounts. Grinding generates close-tolerance dimensions, or produces smooth finishes. Traditionally employed for secondary finishing, grinding is being used increasingly for material removal.

Surface Grinding

Plain surface grinding is accomplished by traversing the workpiece beneath the wheel or by feeding the workpiece (or sometimes the wheel) across perpendicularly to the direction of traverse. The surface being ground is parallel to the axis of the grinding wheel as shown in Figure 4-41.

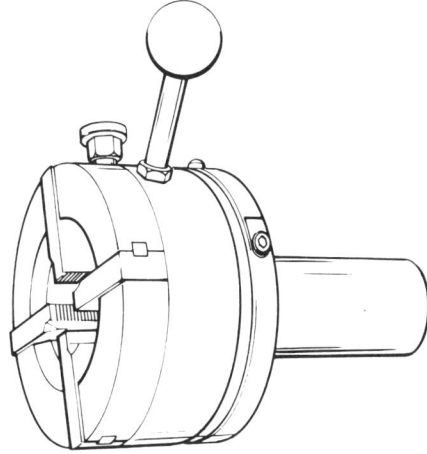

Figure 4-40. *Die head and chasers.*

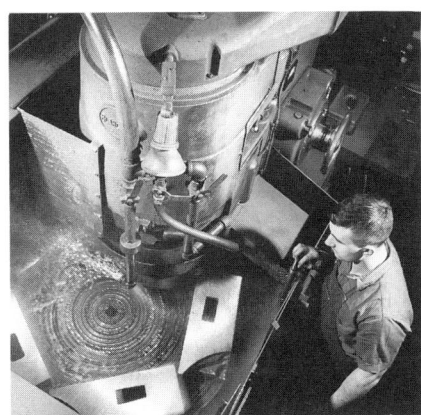

Figure 4-41. *Flat-bed surface grinding operation. (Courtesy, The Blanchard Machine Company)*

Internal Grinding

There are two types of internal grinding processes. In the first, the workpiece reciprocates along the length of the base or cylinder as the wheel rotates rapidly on its axis as in Figure 4-42. The wheel describes circles about the axis of the cylinder being ground.

In the second type, the relative rotation of the bore and grinding wheel is the same as in the first type, except that the workpiece is chucked in a revolving headstock as depicted in Figure 4-43. Both work and wheel revolve on fixed axes. The feed function is assumed by the wheel which cuts radially to predetermined depth as grinding continues.

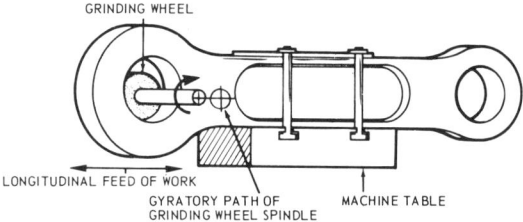

Figure 4-42. *Internal grinding where workpiece reciprocates along the length of the base or cylinder.*

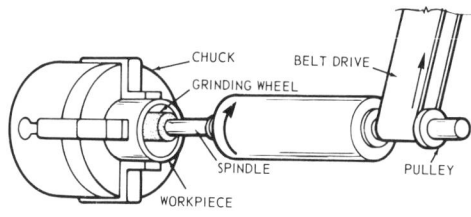

Figure 4-43. *Internal grinding where workpiece is chucked in revolving headstock.*

Centerless Grinding

Centerless grinding, illustrated in Figures 4-44 and 4-45, uses two abrasive wheels. The grinding wheel does the cutting, while a regulating wheel acts as a brake to prevent the workpiece from spinning at the high speed of the grinding wheel. The rotational axis of the regulating wheel is set at a slight angle to the horizontal axis of the cutting wheel to induce horizontal movement of the workpiece. A work rest blade between the wheels keeps the center of rotation of the workpiece above a line between the centers of the regulating and grinding wheels. This ensures a true round cylinder, even though the original workpiece may have been initially out-of-round.

Figure 4-44. *Centerless through-feed grinding. (Courtesy, The Cincinnati Milicron Company)* ⟶

Figure 4-45. *Centerless infeed grinding. (Courtesy, The Cincinnati Milicron Company)*

Cylindrical Grinding

The cylindrical grinding process is used for grinding cylindrical or tapered work on the outside surface only. The abrasive wheel revolves on a horizontal spindle parallel to the work which rotates past the cutting face of the wheel.

Thread Grinding

Thread grinding is essentially cylindrical grinding with a wheel shaped accurately to the thread form to be ground. Threads are ground by means of cutting contact between a rotating grinding wheel plus a relative axial transverse between the two.

Gear Grinding

Gear grinding is especially adapted to finishing gears that require considerable stock removal after hardening, and to produce gears of great accuracy. There are two general classes of gear grinding machines.

Formed Wheel Grinding. This class of gear grinding cuts gear teeth with a contour formed grinding wheel. Three diamonds are used for dressing-one for each side form and one for the outside diameter of the wheel. Accurate head settings must be maintained to obtain consistent tooth profile.

Generation Grinding. This class of gear grinding is similar to hobbing-passing a rack tooth through the space of a revolving gear. There are two general types of generating grinders: one with reciprocating movement between the wheel and gear axially and tangentially; the other with reciprocating movement only in the tangential plane.

GEAR HOBBING

Gear hobbing is a generating process. It consists of a rotating workpiece, a rotating cutting tool (called a hob) with teeth arranged in a helical thread as shown in Figure 4-46, and a machine which maintains a timed relationship between the tool and workpiece. The hobbing machine feeds the tool through or into the workpiece. The cutting edges of the tool have a form that will produce the desired tooth form on the workpiece. The cutting action is continuous in one direction until the blank is finished.

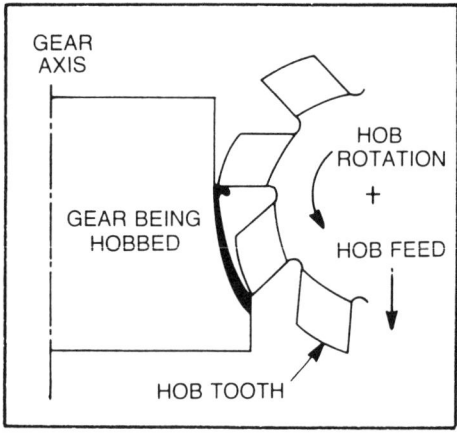

Figure 4-46. *Gear hobbing.*

GEAR CUTTING

Gear cutting is usually associated with bevel-gear manufacture. Some bevel gears are cut on milling machines, but most are of the generated type which are produced on bevel-gear generating machines.

As an example, the usual straight-bevel-gear generator uses two tools simulating teeth of an imaginary crown gear having straighter cutting edges mounted in holders that are, in turn, on a cradle. The cradle, with the tools, rolls during generation in relation to the work spindle on which the gear blank is mounted, as shown in Figure 4-47. The tools reciprocate across the face of the gear blank in a series of cuts as the tools and gear blank roll together, producing the required profile shape of the tooth. The blank is withdrawn at the end of each generating roll, and the cradle and spindle roll back to the starting position for the next tooth.

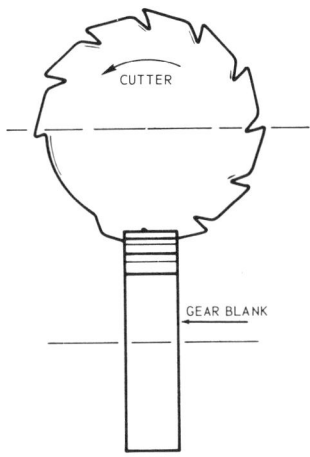
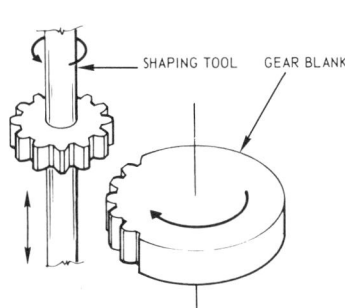

Figure 4-47. *Gear Cutting.* **Figure 4-48.** *Gear Shaping.*

GEAR SHAPING

Gear shaping uses a gear-shaped cutter with relieved cutting edges which rotates and reciprocates; the gear blank rotates but does not reciprocate as depicted in Figure 4-48. The motion of the tool is back and forth across the gear blank, taking a cut with each forward stroke. At the start of the operation, the tool slots into the gear blank through the full depth of one tooth space. Both work and tool are indexed before to each pass until successive passes are made so that cutter and work mesh as two mated gears.

GEAR SHAVING

Gear shaving is one of the finishing operations devised to improve the dimensional accuracy of gears made by hobbing or shaping methods. Both rotary and rack type shaving, seen in Figure 4-49, are used to give such gears a fine, accurate finish before they are hardened.

Rotary Shaving

In rotary shaving, the tool is gear-shaped with teeth that are closely gashed to form many cutting edges. The tool and gear mesh and rotate

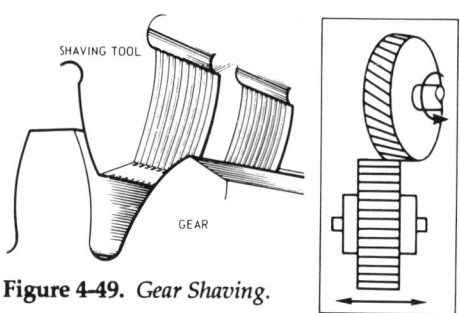

Figure 4-49. *Gear Shaving.*

together. The cutting edges have a sliding action that shears off thin, threadlike chips because the cutter drives the gear being machined, causing their axes of rotation to cross at an angle of 10 to 15 degrees.

Rack Shaving

Rack shaving is essentially the same as rotary shaving except that the tool is a rack that reciprocates in mesh with the workpiece. The shaving effect is obtained by angular motion of the rack or by using helical teeth to finish straight toothed gears, or straight teeth to finish helical toothed gears.

SAWING

Sawing, as a machining operation, uses power-driven equipment to cut material to a desired size or rough shape. Three kinds of machines are in common use in metalworking: the power hack saw, the circular saw, and the band saw. The first consists simply of a saw blade mounted in a reciprocating rack. The machine in operation imitates hand sawing, i.e., the blade cuts on the forward stroke and lifts for each back stroke so that the teeth do not drag in the cut. Cutting speeds are generally low, and feeds generally light.

The circular saw generally replaces the power hack saw for heavy work as illustrated in the multiple saw milling operation in Figure 4-50.

Cutoff band sawing machines were adapted directly from wood-cutting band saws and are used for cutoff of parts that can be handled easily and held on the saw table. They are not feasible for cutoff work on long or heavy barstock.

Figure 4-50. *Multiple circular saw operation. (Courtesy, The Cincinnati Milicron Company)*

ABRASIVE CUTOFF

This process uses an abrasive wheel for the complete cutoff of a piece of material from the workpiece. Almost any type of material can be cut, with high-strength thermal-resistant materials being the most difficult. Abrasive cutting is fast, accurate, and provides good finish characteristics in most materials.

HONING

Honing is an abrasion process to bring both external and internal cylindrical surfaces to a high-quality finish with high-dimensional and geometric accuracy after machining and finish grinding. Abrasives in stick form are spaced evenly about the tool used for internal honing. These maintain an outward pressure against the bore surface to cause the grits to cut, and to provide the feed action as the diameter increases. The motion of the honing tool is twofold: it reciprocates entirely through the length of the bore, and rotates slowly at the same time.

Honing stones are made to specifications similar to those of grinding wheels. Aluminum oxide, silicon carbide, or diamond grits are bound in stick form with a metallic, resinoid, or vitreous bond. Various grades and grains are available.

LAPPING

Lapping is another means of producing a fine finish on ground parts or flat surfaces. Lapping processes usually remove no more than .001 inch (.025 mm) of material. Extremely fine, loose abrasives (called flour) are used to rub down the surface. On ground parts, a tool of relatively soft material such as brass, copper, cast iron, or lead is used. For internal lapping, the tool is designed to expand as the hole enlarges or, for external lapping, to contract as the outside diameter decreases.

REFERENCES

1. MRA Technical Staff, *Machining Data Handbook* (Cincinnati, Ohio: Metcut Research Associates, Inc.,1966) p. 462.
2. R. L. Vaughn and H. B. Miller, "The Why of Metalworking Fluids," *ASTME Paper No. 684* (1965).
3. ADP Machinability Laboratory, Lockheed-California Company, Burbank, California, 1966.
4. C. A. Sluhan, "Cutting Fluids," *ASTME Paper No. 399* (1962), Book 1.
5. B. T. Chao and K. J. Trigger, "Temperature Distribution on the Tool Chip and Tool Work Interface in Metal Cutting," *ASME Paper No. 56-A-87*.
6. M. C. Shaw, "Mechanical Activation-A Newly Developed Chemical Process," *J. Appl. Mech.*, (March, 1948), pp. 37-44.
7. R. L. Vaughn, *Shop Familiarization-Machining Section*, Lockheed-California Company, Burbank, California, 1955.

CHAPTER 5
ACCEPTABILITY OF CUTTING FLUIDS

Cutting fluids have chemical and physical properties which influence their performance under machining conditions, i.e., cutting and grinding operations. Because there are so many variables involved in manufacturing operations (metallurgy, type of cutting tool or grinding wheel, machine type and condition, operator skill, judgement, experience, etc.) any variations in the chemical and physical properties of the cutting or grinding fluids used can be critical. Therefore, *reliability* is a most important criterion of a good metalworking fluid.

To assure *reliability*, many users as well as manufacturers of cutting and grinding fluids establish research laboratory procedures to evaluate these fluids in terms of uniformity of the product, and satisfactory performance in service.

Even though physical, chemical, and mechanical tests are performed on a fluid in the laboratory setting, such tests cannot be used as the only or absolute criteria for fluid quality or performance. Fluid performance should be verified by testing in a clean machine tool under actual production conditions, or in a specially equipped machinability laboratory. Proper evaluation requires that the fluid be run over a period of weeks or months. Such a "test run" can uncover product qualities that cannot be determined by research laboratory evaluation techniques alone.

When cutting or grinding is performed, the physical factors of unit pressure, temperature, and time are involved, together with the chemical factors of the tool and workpiece (or grinding wheel and workpiece). In these operations, the fluid helps to control the built-up edge or reduce diffusion at the tool/chip/workpiece interfaces. No series of research laboratory experiments exists that will duplicate the forces involved under actual plant production conditions. This makes a machinability laboratory very useful.

A machine or machines with a specific series of cutting tools (or grinding wheels) should be set up and run with the fluid under consideration. If a water miscible type is used, the mineral content of the water should be determined and taken into account. In addition, the type of atmosphere present (acidic, high or low humidity, with or without minerals) should be known and its effect should be considered. Thus, how a fluid will perform as a broaching fluid should be evaluated by putting it on the broaching job to be done. The same applies to grinding fluid; it should be tried on the particular grinding operation (e.g., surface, centerless, internal-- they are all different). If the fluid is to be used as an aid to drilling or tapping, etc., it should be tested on those specific jobs.

For example, there is no so-called "lubricity" test or similar test which will give the same answers as a production test. Nor is there a laboratory corrosion test yet devised that will positively predict how a fluid will act in production over a period of time. One of the reasons for this is that the atmosphere in a laboratory usually differs from that in a production shop. Secondly, a fluid in use will usually suffer bacterial degradation. The same fluid tested fresh in the laboratory which may appear to be satisfactory, could actually cause corrosion under production conditions after a week or more in use.

Another factor that should be recognized is that the accuracy of laboratory test results is contingent upon the personal biases and observations of the technician. What one person considers important may not be to another and vice versa. For example, one technician may think that bacterial decomposition is not important. Consequently, production may suffer from loss of the lubricating effects of the fluid in cutting and grinding operations. Further, the corrosion of machines and parts can also result from bacterial attack on the coolant. The natural "stink" from this effect will be noticed by the operator. The operator may start the machine in the morning and walk away from it for 30 minutes or more until most of the odor is released, wasting valuable production time. However, if the laboratory personnel recognized the importance of bacterial resistance, and understood the means possible to control bacterial attack, a much more *reliable* fluid would result.

Many production personnel often reject a particular fluid, not because it does not perform its cutting or grinding functions satisfactorily, but because its use may produce undesirable effects such as: corrosion of machines and/or parts, gummy or hard residues which may damage moving parts on the machine, nauseous odors (stink), and (very infrequently) dermatitis. These undesirable effects may not always be caused by the fluid concentrate. In the case of water miscible types in particular, they can often exist because of the use situation; most frequently because of a poor water source, as explained in Chapter 6.

Research laboratory comparisons are useful as a guide, but how the product holds up under production conditions should be considered the ultimate test. In addition, the physical, chemical, metallurgical, and "mechanical" compatibility of the fluid with the job in terms of cost and quality over "performance" can only be evaluated on the job, in production.

PHYSICAL LABORATORY TEST

Reliability of the product under actual use conditions should be the prime goal of laboratory investigation and control. Laboratory tests should be designed to approach, as nearly as possible, the situations to be met under production conditions. The following tests are examples developed through years of experience by manufacturers and users of cutting and grinding fluids. They are not to be considered standards for acceptability. These tests are only set down as guides and should be considered as examples of possible tests.

Stability

The product should be as stable as possible under normal storage conditions to ensure uniform quality. Some typical tests for stability are:

1. Place a 2 oz. (59 ml) sample of the concentrate in a 4 oz. (118 ml) wide-mouth glass bottle and permit it to remain open to the atmosphere for a specified number of days. If there is any evaporation, phase change, jelling, or solidification, keep the raw material container sealed between withdrawals in the plant.

2. Allow 4 oz. (118 ml) of the concentrate to remain covered in a 4 oz. (118 ml) wide-mouth glass bottle and stand under ambient conditions for a specified number of days (preferably 30 or more). There should be no change such as separation, viscosity, etc.

3. Place the concentrate in a 4 oz. (118 ml) bottle in a cold test chamber at -22°F (-30°C) for a period of 24 hours. Examine for phase separation, flocculation, and sedimentation. If it is a water miscible type (emulsion, semichemical, or chemical), mix it with distilled or deionized water, and also with typical plant coolant water to determine if the concentrate mixes properly after being subjected to this low temperature.

   The temperature of -22°F (-30°C) may never be reached under normal storage conditions. In this case, determine the lowest temperature that the concentrate could experience and test it at that temperature to ascertain whether its function could be impaired.

Other physical tests on straight oil products established by the American Society for Testing Materials are:

Viscosity, Saybolt	ASTM D 88-57
Flash, Cleveland Open Cup	ASTM D 92-78
Fire, Cleveland Open Cup	ASTM D 92-78

Gravity, A.P.I. ASTM D 287-82
Foaming ASTM D 892-74
Corrosion, Copper Strip ASTM D 130-83
Emulsion Stability ASTM D 1479-57T

Modifications of the ASTM Emulsion Stability test may be run as follows:

4. To determine stability and effects of highly mineralized water, prepare 3% mixtures of each of three products to be tested, two in distilled water and one in synthetic hard water which contains 1,000 parts per million (.1 mole calcium chloride in 2.6 gallons [10 liters] of distilled water) hardness. Allow one of the distilled water samples to stand quiescent for 24 hours at room temperature, keeping the other at 200°F (93.3°C) in a constant temperature bath for 24 hours. Allow the sample made with the hard water to stand for 24 hours at room temperature. Ratings of degree of separation or turbidity can be obtained after 24 hours using Table 5-1 which is scaled from zero to six.

5. Place the test emulsions at the prescribed concentrations in 4 oz. (113 g) oil sample bottles and allow to stand at ambient temperature for 24 hours. Examine for oil separation, emulsion layering, and sedimentation.

Table 5-1. Ratings of Degree of Separation or Turbidity

Rating	Separation or Turbidity
0	None
1	Very slight
2	Slight
3	Slight to moderate
4	Moderate
5	Moderate to complete
6	Complete

Residue

Nonfluid residues left after water has evaporated from water miscible fluids *can* interfere with some machine tools but not with others. As machines become more fully automated (e.g., transfer, CNC) it is more necessary than ever that machines operate without impairment. The ideal residue would be one which is fluid, has high lubricating value, and is easily water resoluble. Some methods for testing are:

1. Put 0.17 oz. (5 g) of concentrate in a 0.50 by 2 in. (13 by 50.8 mm) glass petri dish and allow it to stand in the room for 30 days to one year.

2. Repeat, putting concentrate into a desiccator and letting stand for one month, replacing the spent dehumidifying material as often as required to remove all moisture.
3. Repeat, putting material in an oven at 221°F (105°C) for 24 hours. Examine for tackiness, crystallization, solidification.

Paint

Some paints used on machine tools can be "lifted" or disintegrated by cutting and grinding fluids, mostly by the water miscible types. This can create a problem because, if the dissolved paint recirculates through the coolant system and then lies between machine moving parts, the parts can "stick" and not function. Water miscible type fluids, high in wetting power, are most prone to penetrate and "lift" machine tool paints of unsuitable composition or improper application. The high alkalinity of water type coolants can also hasten the paint dissolving action.

Paints are available which will resist the lifting and dissolving action of water miscible fluids. Since any type of paint requires proper machine tool surface preparation, the use of the correct priming coat and the correct final coat is elementary.

A simple test for paint resistance is as follows:

Make 10% mixtures of the fluid concentrates in distilled water. "Soak" the steel panels (which have been properly prepared and painted) in the coolant solutions (room temperature) placed in suitably sized covered jars. One side of the panel should be left intact, the other "scored" (with a nail or other suitable tool) to the bare metal. Observe the panels every month for 6 months to note any "lifting," dissolving, or color change of the paint film.

CHEMICAL LABORATORY TESTS

Putrefaction

Resistance to bacterial attack is one of the most important requirements for reliable performance of water miscible fluids in production. Bacterial attack on a fluid reduces its cutting and grinding lubrication functions. It also promotes gumming and clogged coolant lines, and induces corrosion and stink.

Simple laboratory tests for bacterial presence can be conducted using the widely accepted Rossmore Protocols. These protocols take six weeks.

After the six-week bacterial growth studies, make corrosion tests on 1 by 2 in. (25.4 by 50.8 mm) rounds, or 0.12 by 1 in. (3.1 by 25.4 mm) squares, or both (see test under "Steel Corrosion"). It is also desirable to run pH tests on the coolants undergoing the bacterial-aeration growth studies on a weekly basis.

Water Compatibility

The following steps should be followed:
1. Make solutions of 1%, 4%, and 10% concentrate in the water to be tested.
2. Put into 3.3 fl. oz. (100 ml) stoppered glass graduates and let stand 24 hours.
3. Repeat with distilled water and compare.
4. Repeat, using plant water which has been evaporated four times (and replenished after each evaporation) to note the effect of the concentration of solids simulating production conditions. If the fluid is chemical, observe for solids effects, clarity, foam, and corrosion. Test for corrosion on steel and cast iron rounds or squares.
5. For emulsion fluids, check for emulsion stability, creaming, separation, foaming, and corrosion on steel rounds or squares.

pH

Make 1% and 4% mixtures in distilled water and run a pH test, using a glass electrode. The pH should be between 8.0 and 10.5 for most situations.

Nonferrous Corrosion

Nonferrous metals being machined or ground should not be corroded. Unfortunately, some of the most effective ferrous corrosion inhibitors will readily attack certain nonferrous metals. Further, because some fluid plumbing systems in machine tools (valves, pipes, etc.) are nonferrous, it is important to test fluids for their potential corrosive effect on various metals, either alone or in combinations. Some procedures are:
1. Make a 10% mixture of the coolant in distilled water. Fill 19 test tubes half full. Properly abrade and clean .030 by 0.25 by 4 in. (0.76 by 63 by 102 mm) strips of copper, yellow brass, aluminum (7071), galvanized steel, and cold rolled steel. Put single pieces of metal in the test tubes, then pairs, then three, then four, finally all five pieces together. (Specific alloys under investigation can replace those cited.) Stopper the tubes and observe reactions daily for 10 days. After 40 days, remove the strips from the tubes

and place in dry tubes. Stopper the solutions and keep as a permanent record if desired.
2. Make up a galvanic cell of a 1 by 6 in. (25.4 by 152.4 mm) panel of 2014 T-6 aluminum (or other desired practical metal) and a panel of 1020 cold rolled steel. Couple these panels with a piece of aluminum wire and separate the upper portion by a piece of glass tubing. Immerse approximately 3/4 of the bottom portion of the panel in the test fluid for 48 hours. Examine the panels for salt growth, corrosion, and discoloration. Also note the condition of the fluid after the immersion period.

These tests can be repeated using typical plant coolant water, as well as the same water evaporated four times (and replenished after each evaporation) to determine the effects of water solids buildup due to water evaporation.

Cast Iron Corrosion

Many manufacturers are using water miscible fluids for machining cast iron and similar irons in order to improve tool life and control size at higher rates of production. Some procedures for testing these fluids and their resistance to corrosion are:

1. Place screened, washed, and dried cast iron chips in a plastic petri dish which has a sheet of filter paper on the bottom. Pour the test fluid over the chips and allow to soak for a period of 30 minutes. After this period of time, decant the fluid and partially cover the chips with the petri dish cover. After 24 hours, examine the chips for corrosion, and rate them according to the amount of corrosion visible on the filter paper.
2. Make 1%, 2%, 3%, 4%, and 5% mixtures of the concentrate in distilled water. Place 0.35 oz. (10 g) of clean cast iron chips (made from a uniform cast iron and machined without any coolant) into five, 1.7 fl. oz. (50 ml) beakers. Into each beaker place 0.07 fl. oz. (2 ml) of the respective 1%, 2%, etc. solutions. Mix each thoroughly, then place each separately onto a piece of filter paper inside a 0.59 by 2.36 in. (15 by 60 mm) petri dish. Allow to stand for 24 hours under room conditions and observe any indication of rust.
3. Wash malleable iron chips in petroleum ether to remove all traces of oil or other contaminants which might affect the corrosion test. Place 0.07 oz. (2 g) of washed chips in a 2 in. (50.8 mm) square area of a clean, polished, and ground case iron plate. Pipette 1 cc. of the test fluid over the chips on the plate. The chips should not rust nor should the plate stain or rust in 24 hours, when the emulsion concentration is 2% or over.

These tests can be repeated, using plant coolant water and the same water evaporated four times (and replenished after each evaporation) to determine the effect of buildup of water solids on corrosion.

Steel Corrosion

No fluids should corrode (rust) metals being machined or ground, nor should they corrode the machine tools. Unfortunately, laboratory ferrous corrosion tests cannot always predict production results, but they are a useful guide. Figure 5-1 relates the possible chemical reactions of cutting

ROOM TEMPERATURE	CUTTING ZONE TEMPERATURE
NO (will not corrode)	NO (will not work as cutting fluid)
DESIRED → NO (no corrosion)	YES (works as cutting fluid)
YES (will corrode)	YES (works as cutting fluid)

Figure 5-1. *Possible chemical reactions of cutting and grinding fluids.*

and grinding fluids at room and cutting zone temperatures to corrosion and cutting fluid action. The most desirable condition for efficient, noncorrosive fluid action is singled out. The breakdown of a fluid by bacteria while in use, the concentration of water solids by evaporation of water while the fluid is in use, and the changes of constituents in the atmosphere, can all make an otherwise satisfactory fluid cause corrosion on parts in production.

Some steel corrosion tests in use are:

1. Immerse a strip of 1020 cold rolled steel in the test fluid of 1%, 2%, 3%, 4%, and 5% mixtures of concentrate for a period of 48 hours. After this time has elapsed, examine the steel for signs of corrosion and discoloration.

2. Make 1%, 2%, 3%, 4%, and 5% mixtures of the coolant concentrate in distilled water. Dip ground 1020 steel round rods, 1 by 2 in. (25.4 by 50.8 mm), in the respective solutions and build a three-piece pyramid. Place the pyramids in five sufficiently large petri dishes and let stand under room conditions for 24 hours. Observe for rust or stain.

   Using the same concentration solutions, run this corrosion test on one inch square by 0.12 in. (25.4 mm^2 by 3.1 mm) thick 1020 steel plates. Dip the cleaned and abraded pieces in the solutions, stack two pieces together on a watch glass, and let stand for 24 hours under room conditions. Observe for rust or stain after separation.

3. Repeat step 2, putting the 1 by 1 in. (25.4 by 25.4 mm) plates onto a cleaned and abraded cast iron plate surface. Let stand for 24 hours and observe for rust before and after removing the steel squares.

Repeat the above, using plant coolant water and the same source water which has been evaporated four times (and replenished after each evaporation). Note the effect of water solids buildup from evaporation of the water under production conditions.

Machinability

At the present time, there is no standard procedure for testing the machinability characteristics of cutting and grinding fluids. However, to compare the effect of particular cutting fluids on actual machining operations, run *controlled* performance tests--using several materials--on all types of operations of interest, such as:

1. Single point turning
   Measure and compare depth of cut, chip load, feeds, and speeds.
2. Drilling
   Measure and compare size of hole, feeds, and speeds, and depth of cut.
3. Tapping
   Measure and compare hole size, length of thread, feeds and speeds, blind hole or open hole.
4. Milling
   Measure and compare depth of cut, chip load, and feeds and speeds.

By using a controlled test on all operations, where the only variable is the fluid being tested, a true evaluation of the effect of the test fluids on machinability should result. Size control, finishes, and tool life would be the criteria for the relative effectiveness of the fluids.

MECHANICAL LABORATORY TESTS

Foaming

Excessive foam can be objectionable. Some fluids high in wetting agent content or having high "wettability" will tend to foam more than those which have lower wettability or higher surface tension. Machines which "whip" air into fluids (1) and do not have sufficient sump capacity for foam subsidence will cause more foam to be generated. Laboratory tests for foaming cannot be used to predict accurately how a fluid will foam in a particular machine tool, but may give an indication. Some possible foam tests are:

1. Place 3.3 fl. oz. (100 ml) of fluid (emulsion, chemical or semi-chemical) of 1%, 2%, and 4% concentration into a 6.76 fl. oz. (200 ml) glass-stoppered cylinder. Agitate the cylinder vigorously for one minute and time the period required for the foam to break. A time interval of up to 30 seconds is acceptable.
2. Place 7 fl. oz. (503 ml) of water in a 27 fl. oz. (800 ml) Griffin beaker and heat to 120° F (48.9° C). Stir the water at 1,300 rpm while slowly adding 0.8 fl. oz. (24 ml) of the concentrate. Stir for five minutes. If the foam goes over the top of the beaker while agitating, the product has failed. If, after five minutes, the foam has not gone over the top, stop the agitation and measure the foam height initially and after standing for 10 minutes. Record and compare the foam heights of the various fluids tested.

A modification of this test is to use an electric "blender" (1) and stir a specific volume of the fluid of 1%, 2%, and 4% concentrations for a specified number of seconds. Compare foam heights and time for the foam to "decay."

Dropout Rate

Some systems require foam to carry fines in an overflow cleaning system while other systems require a minimum of foam for operation. Once the fines reach their separation point, certain systems depend upon some form of gravity (natural or induced, as in a continuous centrifuge) to settle these fines. An indication of the performance of a new fluid may be obtained by mixing a known quantity of fines into a known sample and timing its dropout rate using a graduated cylinder or Imhof cone.

Tramp Oil

By deliberately contaminating fluid with hydraulic and lubricating compounds, the following can be determined during performance testing:
1. Whether leaks in hydraulic and lubricating systems may be the cause of difficulties with a particular cutting or grinding fluid
2. Whether tramp oil is emulsified by products in this category
3. Whether this causes a drop in performance
4. Whether filtering is hampered by tramp oil

Oil Reclamation

Mixing emulsifiable and neat oil products, plus tramp oils, into chip handling systems designed for oil reclamation may prove troublesome. Excess emulsifiers fed into such systems can reduce the reclamation rate to nearly zero if handled together. Separate handling and reclamation systems, or washing of chips, may be the only solution to this problem.

METALLURGICAL AND CHEMICAL COMPATIBILITY (2)

All metal cutting processes involve the removal of material normally having a protective covering (film). The removal process also includes the exposure of unprotected, unfilmed surfaces to a variety of chemical elements in the cutting fluids and in the atmosphere. This exposure occurs first under conditions where temperatures and pressures are elevated for various time periods, depending on the process. In these cases, the finished workpiece is normally set aside for various periods (sometimes days or weeks). During this time, varying amounts of cutting fluid cover all newly-produced surfaces in environments ranging in humidity and temperature. If they are compatible, no unwanted, detrimental or unknown reaction will take place between the workpiece and its environment. Unfortunately, the determination of possible chemical interaction, i.e., corrosion, depends upon the specific environment and workpiece.

The effects of *residual films of cutting lubricants* on the metallurgical integrity of metallic parts operating under conditions of high stress and temperature has become a controversial issue, especially among producers of aerospace products. This concern has been generated primarily by laboratory and service failures of critical components, attributed to stress corrosion crackings and intergranular corrosion. In general, failure analyses of components and laboratory studies have associated these failures with the presence, in the service environment, of chloride and other ions of the halogen family-iodine, bromine, and fluorine, as well as phosphorus or sulfide ions. Limited information is available in the published literature, however, pertaining to the relationship between these ionized chemicals and specific metalworking fluids.

It has been established in a large number of laboratory studies that the presence of metallic salts of chlorine, principally NaCl and KCl, will produce stress corrosion cracking in material such as high-strength steels, austenitic stainless steels, and titanium alloys when tested in high-temperature/high-stress environments. These studies, however, have not been related to many actual metalworking fluids, although these elements are common in today's cutting fluids. In one investigation (3), which established that the solvent trichloroethylene (CHCl:CCl) would produce stress corrosion cracking in the titanium alloy Ti-5Al-25Sn(A-110), a cutting oil tested in the same program did not produce cracking. However, an Air Force sponsored study (21) concluded that the use of chlorinated and sulfurized cutting fluids showed no difference in the mechanical properties of selected aerospace alloys versus the use of nonchlorinated/sulfurized cutting fluids.

In essence, there is a lack of fundamental data related to physico-chemical reactions which occur between metal surfaces and the chemical constituents of metalworking fluids in high-temperature/high-stress environments. Basic metal cutting research (4) has shown that certain chemical substances will react with specific compositions of workpiece and tool materials to form metallic compounds of low shear strength in metal removal operations. These studies, however, have never been correlated with stress corrosion or other metallurgical effects.

Other basic research of highly theoretical nature (5) has shown that surface active agents such as stearic acid and oleic acid, which are basic ingredients of many metalworking lubricants, will produce metallic soaps which are adsorbed on metal surfaces and can produce a lowering of mechanical strength in these surface layers. These data generally cannot be used as guidelines for practical metal cutting lubricant selection. They are, however, indicative of possible deleterious metallurgical effects that can be produced by various chemically active metalworking fluids. The possibility of these effects has been established. The basic parameters associated with this occurrence, however, have not been determined. It is recommended that *caution* be observed in the selection and application of fluids in the processing of hardware which must operate within a critical range of stress and temperature.

Cleaning procedures can be established which, if carefully followed, will remove all traces of metalworking fluids from surfaces of machined parts. These procedures followed under 100% compliance are entirely satisfactory for parts not containing internal recesses or crevices in which fluids can be trapped, and which cannot be adequately cleaned. For critical components which cannot be thoroughly cleaned, or where doubt exists over the effects or residual contamination of fluids, or where total procedure compliance may be too uncertain or costly, the only reliable approach is to perform simulated service testing, testing the part material by putting it in the presence of the fluid to which it will be subjected during processing. Elimination of questionable fluids from use on the production floor may often prove the best insurance against possible stress corrosion cracking.

General Corrosion Types

If the metal and cutting fluid are compatible and direct chemical reaction is not involved, there are still the problems imposed by localized attack which can frequently be severe.

Staining is a form of localized attack. It covers relatively large areas, and can be classified as superficial and self-inhibiting. The process will prevent further corrosion in most cases. *Rusting, spotting,* and *pitting* are more concentrated attacks in a smaller area. Rusting is more destructive than staining to the normal workpiece.

Stress Corrosion

No commercial metals or alloys are perfectly homogeneous but contain inclusions, impurities, banding, and segregation of chemical compounds in grain boundaries and in the grain. Within the grain structure, the crystals (composed of atoms in lattice patterns) also present a nonhomogeneous structure, with voids or pits of molecular size, adsorbed impurities, and steps and dislocations of the lattice. When corrosion of any type is combined with stresses inherent or imposed in the surface or atomic structure, the combined action is termed stress corrosion or, referring to the result, stress corrosion cracking.

This particularly insidious form of corrosion does not oxidize large amounts of metal. Instead, it may preferentially attack grain boundaries or natural and ever present dislocations in the crystalline structure of the grain itself. Catastrophic failures have occurred in aircraft, chemical plants, and boilers under stress levels as low as 1% of the design intent. Insuring against possible initiation of such corrosion action is necessarily the concern of the entire machining industry. The halogen group of chemicals all have produced stress corrosion cracks in titanium alloys when subjected to stress at elevated temperatures during laboratory tests.

Certain factors appear conducive to rapid formation of stress corrosion:

1. Planar dislocations extending through the protective surface film to form new unprotected slip steps. These are produced continuously in any metal removal or metal forming process.
2. Rapid exposure to an environment containing chemical elements corrosive to the material which produces the initial pit that can later develop into a tunnel.
3. Motion of the dislocations under the action of the applied stress which continues the growth of the corrosive pit.
4. High temperatures which create increased energy levels that accelerate these actions.
5. The exposed positive metal ions attract the negatively charged ions in the environment. The major members of the halogen family--chlorine, bromine, iodine and fluorine, as well as sulfur and phosphorus--can provide such negative ions and corrosive properties. Many of these elements are commonly found in conventional fluids.

Embrittlement

Another stress corrosion form, peculiar to certain high-strength steels, which may cause failure through the diffusion of hydrogen ions or atoms into the metal, is termed hydrogen embrittlement. Stressed steels generally suffer caustic embrittlement in the presence of hot caustic soda concentrations. Similar brittleness occurs with certain gold alloys upon exposure to solutions containing chloride ions. Historically, formed brass has been noted for season cracking under environments containing ammonia traces.

Elemental Diffusion

Elemental diffusion is another mechanism of stress corrosion which may occur in the presence of iodine and other lubricants. At certain cutting speeds and feeds, the cutter edge/surface pressures create optimum conditions for elemental diffusion. The diffused halogen or other elements in the presence of an electrolyte will develop not only pit corrosion but, under stress conditions, stress corrosion cracking. Electrolyte may be produced by moisture and many kinds of surface contaminants, such as salts. Metals are not alone in this regard; the exposure of stressed methacrylate resin plastic to acetone produces cracks and failure.

While there are many causes of stress corrosion cracking, it is difficult to predict when this type of failure will actually occur. Each alloy is susceptible to only a few rather specific environments. The stress conditions, structure, composition, heat treatment of the alloy, and duration of exposure all have a bearing on the cracking (6).

For further studies and documentation in the general area of stress corrosion the reader is directed to References (7) through (20).

Corrosion of Machines

Oil emulsions are usually not corrosive to machines unless the chloride and sulfate content of the diluting water is high. If iodine lubricants are used, corrosion can be a serious problem. There is little need to discriminate between rust and pitting corrosion of machines since these are the same kind of corrosion, although pitting partially and locally inhibits rust. Local corrosion can occur in the case of dissimilar metal contact areas, i.e., aluminum versus steel or cast iron, and magnesium versus steel.

Metallic deposits occur in rare cases from galvanic processes where the machine is the cathode and another metal acts as an anode in the presence of an electrolyte.

When machining is interrupted and time allows the cutting fluids to dry, a residue may remain attached to machine surfaces with certain fluid types, mainly chemical.

Machine lubricants also deteriorate emulsions by gradually amalgamating suspended oil globules. The oil separating from the emulsion, plus the lubricating oil, collects on the surface of the cutting fluid tank, sealing the cutting fluid from aeration. The oil-sealed fluid provides an ideal condition for anaerobic bacterial growth (see Chapter 6). The tremendous mitosis of anaerobic bacteria chemically decomposes the cutting fluids. Decomposition product residues are then found on machine and workpiece.

HUMAN COMPATIBILITY

After physical, chemical, and mechanical compatibilities are satisfied, the fluid must still be accepted by the operating personnel. Appearance, "feel" as to slipperiness, ability to see the work being cut or ground, and other "nonperformance" characteristics may influence the acceptance or rejection of a particular fluid. Inherent product odor or the odors caused by bacterial action may be further determinants as to whether a fluid "works" or not.

Above all, no fluids should be used if they are harmful to personnel health. Careful screening of raw materials by the fluid manufacturer is normal practice, and finished products are usually tested by competent dermatological laboratories or impartial institutions for evaluation before the products are widely marketed. As a result, most cutting and grinding fluids in use today are quite safe for normal industrial applications.

Dermatitis

Human beings are subjected to all kinds of work situations and individual reactions vary. Consequently, it is desirable to list a number of precautions that should be exercised to prevent, or reduce as much as possible, the number of susceptible people from experiencing dermatitis. Dermatitis is defined as an inflammation of the skin. It is usually caused by:

1. Use of solvents
2. Some cutting oils which irritate the skin
3. Cutting oils which break down in the presence of water
4. Too high an alkalinity
5. Too high an acidity
6. Too "rich" a mixture of cutting fluid

7. Use of improper measuring and mixing equipment
8. Use of some "protective" creams
9. Some metals such as chromates, zinc, cadmium and more
10. Use of improper microbicide or misuse of a satisfactory microbicide. It should be noted that microbes include bacteria, fungus and yeasts.

Most people who use metalworking fluids normally have their hands in contact with the fluid many hours of a working day. In contrast, total time spent washing the hands might be less than 10 minutes per day.

Typically it is the lighter complexioned personnel who have the "driest" skin which is most sensitive to any exterior environment.

The skin has a pH of 6.8 (on the acid side) and has a protective layer of natural oils to retard moisture evaporation and to act as a mechanical shield. Consequently, anything which tends to remove the natural oils from the skin and to neutralize the acidity will tend to cause trouble, or produce dermatitis.

Personnel will experience severe reddening of the skin when working with solvents or low boiling organic liquids. Kerosene, naphtha, mineral seal oil, and chlorinated solvents are materials which remove the natural oils from the skin. This permits moisture to evaporate from the skin and will usually result in a dry condition and cracking of the skin, particularly between the fingers. It also leaves the skin more open to attack by anything irritating to it.

If the use of such materials cannot be avoided, proper protective gloves or a water soluble protective cream should be used. It is important which cream is used, because, often, these creams have microbicides in them and the microbiocides themselves can be irritants.

Dermatitis is seldom caused by bacteria, in spite of widespread opinion to the contrary. Most dermatitis is the result of chemical and physical effects. In addition, a person can use a material for a period of time without difficulty and suddenly become reactive to it. This is called "building up a sensitivity." Or, they may be sensitive to a product for a period of time and then build up an immunity to it.

Certain chlorinated sulfurized or chlorinated cutting oils react with the skin to produce chloracne. Some of these oils can also break down with moisture of the skin to form mineral acids which also cause skin irritation.

Dermatitis can occur with certain types of (but not all) chlorinated sulfurized cutting oils when they get into a water miscible fluid. The water will gradually break down the chlorine in such cutting oils and release hydrochloric acid. This does not occur in the first five minutes or so, but occurs over a period of a few days or a week or more. The important point is that, as the chlorine is broken down the concentration of acid becomes higher, and as it does, it helps to break down increasing amounts of the chlorinated oil more quickly. This explains why chlorinated oil could be mixed with a water miscible coolant with no trouble occurring in the first day or the first week or more of use. However, as the concentration of the acid builds up, a point is reached where it has a very irritating effect upon the skin and this would be called a "primary irritant." Almost anyone could experience an adverse reaction on the skin from such a situation. A further result would be corrosion of machine tools and machined parts.

There are a number of situations wherein chlorinated or chlorinated sulfurized cutting oils can become mixed with water miscible fluids. The first case would be where a machine has had a particular chlorinated sulfurized cutting oil in it which has been replaced with a water miscible product. If the machine is not thoroughly cleaned, enough oil is left in pumps, pipes, and other areas of the machine, as well as on exposed surfaces, to be washed off and mixed with the water miscible product. The machine, if at all possible, should be flushed out with a petroleum solvent or cleaned with an alkaline cleaner. Ideally the first charge of water miscible fluid should be added, run from three days to a week, then discarded. A fresh charge is added which should be capable of running satisfactorily for a normal period of time.

The second way cutting oils can get into the water miscible solution is if the operator is running it in a lathe, using it for turning, drilling, boring, and facing. If the metal is of a type that a straight oil must be used to get the proper finishes, size control, and chaser or tap life, the operator would normally brush a straight cutting oil onto these reaming, tapping, or threading operations. In this instance, the water miscible fluid would surely come out of the spigot occasionally, wash onto the part, and mix with the cutting oil.

A third way is where a cutting oil is used in a primary operation, following which the part goes into a secondary operation where a water miscible fluid is used. In this way, oil would carry over on the part into the secondary operation and again be mixed with the water miscible fluid.

The fourth way is when an operator running a particular chlorinated sulfurized cutting oil, assumes that the machine is thoroughly cleaned, does not use any protection cream or gloves, and gets his hands into the water miscible fluid during normal working motions. In this case, the operator's skin has been so saturated with the cutting oil that it takes from one to two weeks for this oil to gradually work out of his skin. In the meantime, if the operator is working in water miscible fluid, the water in the mixture can react with the oil in the skin and again produce acid followed by irritation. To alleviate the situation, the operator should use a waterproof barrier cream or waterproof gloves until the oil has a chance to work out of his skin (one to two weeks).

The worst situation for the development of dermatitis in most shops, is where a set-up man or operator will go from a machine using cutting oil to one using a water miscible fluid and use a solvent for washing off parts between these operations. The solvent will remove the oil from the skin and leave it exposed, so that if the cutting oil does break down in the presence of water, it can really *attack* the skin.

Here, protective creams are of no value. To prevent attack of oils on the skin, use a cream that is water soluble, but oil insoluble. To prevent the attack of water miscible products on the skin, a water insoluble or oil soluble product must be used. If an operator's hands are in an oil, then solvent, then water miscible fluid, no barrier or hand cream will hold up. In this case, an operator will have to wear protective gloves.

Another cause of dermatitis can be the use of water miscible fluids of too high an alkalinity. The term pH describes the degree of acidity or alkalinity. The scale is as follows:

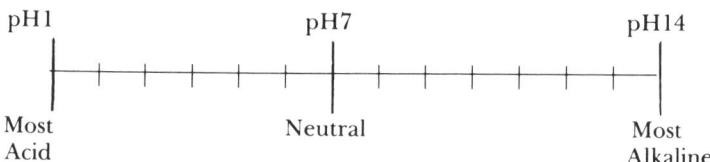

In essence, pH is the concentration of hydrogen (acid) or hydroxyl (alkaline) ions and is represented by the logarithmic pH scale. Water normally has a neutral pH of 7. A solution with a pH of 8 has 10 times as many hydroxyl ions as a solution of pH7, i.e., it is 10 times as alkaline. A solution of pH6 has 10 times as many hydrogen ions as a neutral (pH7) solution or is 10 times as acid, etc.

For better rust inhibition on ferrous metals, and to keep most bacteria under control, the more alkaline the solution (the higher the pH) the better. However, many nonferrous metals are badly corroded by high pH fluids which can also produce dermatitis.

On an operation where the operator's hands do not get into the coolant too frequently or for long periods of time, such as on a surface grinder, a coolant of high alkalinity (high pH) could be tolerated. However, on a throughfeed centerless grinder or a turret lathe, where the operator's hand is in the fluid almost continually, a fluid of high alkalinity can very rapidly counteract the normal acid (pH6.8) condition of the skin and cause dermatitis. In general, a pH greater than 9.0 should be avoided and should actually be the range of pH8.5 for best results.

If a person is sensitive to a pH of 8.5 or more, it is important to keep the fluid at the correct concentration to keep the pH in proper range. A water-resistant cream or protective gloves may be necessary. The cream must be chosen carefully because some creams may contain microbicides which can themselves be irritants.

The presence of excess wetting agents or emulsifying agents may cause dermatitis by drying the skin through removal of the natural oils. The fluids must be kept at proper concentration and, if necessary, a proper barrier or protective cream or suitable gloves should be used.

Dermatitis can be caused by using galvanized measuring and mixing equipment, or using newly galvanized piping and splashguards on machine tools. Galvanized metal (zinc plated steel) *can* react with certain coolants to produce zinc salts or chemical reaction products and produce dermatitis in some people. Therefore, to be on the safe side, all measuring and mixing equipment should be made from plain steel or plastic.

Another source of dermatitis can be microbicides added to the coolant concentrate by the manufacturer, or to the water miscible fluid by the user. Very few of the available microbicides are effective in preventing the growth of bacteria; in fact, many can actually feed the bacteria under the conditions present in coolant usage. Even more important, microbicides can be very irritating to an operator's skin. In general, no microbicides should be added to the coolant without first consulting the coolant supplier. Among other reasons for doing this, many microbicides can be antagonistic to each other, so if the improper ones are mixed, their desired benefits will not be realized.

Dermatitis can also be caused by some metals, metallic salts, or strong oxidizing or reducing agents. Sometimes chromates or cadmium salts are not properly rinsed from plated parts before the parts go to grinders. When this happens, the operator can develop a severe case of dermatitis. All plated parts must be properly rinsed before grinding. In addition, there are a few fluids (usually used for surface grinding) into which manufacturers put chromates as microbicides and corrosion inhibitors. These fluids can sometimes cause dermatitis. One of the best ways to control bacteria growth is to use an alkaline microbicidal cleaner to clean machine tools properly, rinse well after the cleaning, then install a fresh coolant mixture.

Further discussion of the causes and prevention of dermatitis can be found in References (23) through (29).

Possible Toxicity of Certain Chemicals

Such materials as carbontetrachloride, trichloroethylene, alkali nitrites and nitrates, iodine compounds, and many other chemicals have been suggested for use in metal removal operations. Even if some of these chemicals have been previously used with success, they should be examined for possible toxic effects. Two excellent reference sources are "Patty's Industrial Hygiene and Toxicology" (Clayton & Clayton Editors), and "Dangerous Properties of Industrial Materials," by H. I. Sax.

Sometimes a satisfactory operation cannot be performed without the use of what may be considered a "dangerous" chemical; such a chemical could possibly be used with proper safeguards.

REFERENCES

1. A.B. Myler, "Tests for Soluble Oils," *American Machinist* (June 2, 1958).
2. ADP Machinability Laboratory, Lockheed-California Company, Burbank, California, 1966.

3. Defense Metals Information Center, "Stress Corrosion Cracking in Ti-5Al-2.5Sn," DMIC Memorandum No. 60, Battelle Memorial Institute, 1959.

4. H.W. Babel, et al., "An Evaluation of the Present Understanding of Metal Cutting," Battelle Memorial Institute--ASTME Research Fund (1959), pp. 135-189.

5. I. Krames and L.J. Demer, *Effects of Environment on Mechanical Properties of Metals*, Vol. 9 of *Progress in Materials Science* (New York: Pergamon Press, 1961).

6. F.L. LaQue and H.R. Capson, *Corrosion Resistance of Metals and Alloys* (New York: Reinhold Publishing Corporation, 1963), p. 31.

7. C.B. Ward, "Corrosion in Light Alloys," *Light Metal Age* (February, 1966), pp. 16-18.

8. P.R. Swann, "Stress Corrosion Failure," *Scientific American* (February, 1966), pp. 71-81.

9. U.S. Department of the Air Force, "Corrosion Prevention/Control Program," *AFR 400-44* (July, 1964).

10. H. Suss (GE), "Stress Corrosion-Causes and Cures," *Material Design Engineering*, 61, No. 4 (April, 1965), p. 102.

11. R.A. Pride and J.M. Woodward, "Salt-Stress-Corrosion Cracking of Residually Stressed Ti-8Al-1Mo-IV Brake-Formed Sheet at $550°F$ ($561°K$)," *NASA Technical Memorandum X-1082*, Langley Research Center, VA.

12. D.N. Braski and G.J. Heimerle, "The Relative Susceptibility of Four Commercial Titanium Alloys to Salt Stress Corrosion at $550°F$," *NASA Technical Note D-2011*, Langley Research Center, VA.

13. H.B. Dexter, "Salt Stress Corrosion of Residually Stressed Ti-8Al-1Mo-IV Alloy Sheet after Exposure at Elevated Temperatures," *NASA Technical Note D-3299*, Langley Research Center, VA.

14. W.K. Boyd and F.W. Fink, "The Phenomenon of Hot-Salt Stress-Corrosion Cracking of Titanium Alloys," *NASA Cr-117*.

15. E.J. Duffy and D.E. Trout, "Relative Susceptibility of Copper-Nickel Alloys to Intergranular Cracking in Various Media and at Varying Stress Levels," *ASME Paper 65-WA/CT-1*.

16. R.A. Burton and J.A. Russell, "Lubricant Effects on Fatigue in a Stationary Concentrated Contact under Vibratory Loading," *ASME paper 65-WA/CF-3*.

17. G. D. Galvin and H. Naylor, "Effect of Lubricants on the Fatigue of Steel and Other Metals," *IME* (1965), Shell Research, Ltd.

18. R.A. Gorski, "Comparison of Inhibited Methyl Chloroform Solvent to Trichlorotrifluoroethane Solvent," *Freon Lab Report KSS-5327*, E.I. DuPont de Nemours & Co.

19. Anon., "Stability Tests of Solvent/Metal Systems," *Freon Lab Report KSS-4761*, E.I. Du Pont de Nemours & Co.

20. Anon., "The Effect of Organic Solvents and Sodium Chloride on Titanium Alloys," *Freon Lab Report KSS-5435-B*, E.I Du Pont de Nemours & Co.

21. N. Zlatin, J.D. Chestopher, J. T. Cammett, et al., Metcut Research Associate, Air Force Materials Laboratory Report, AFML-TR-73-165.

22. H. Simon, M. Thomas and Knut Maier, "WT-Zeitschrift Für Industrielle Fertigung" (Springer-Verlag, 1979).

23. W.L. Lea, M.D., "Cutting Oil Dermatitis," *Wisconsin Safety News* (October-December, 1955), pp. 10-11.

24. _____, "'Soluble' Cutting Oils Discussed," *Wisconsin Safety News* (July-September, 1955), pp. 10-11.
25. L. Schwartz, M.D., *The Prevention of Occupational Skin Disease* (New York: McGraw-Hill, Inc., 1955) Sponsored by the Association of American Soap and Glycerine Producers.
26. S.H. Osborn, *Facts About Oil Dermatitis* (I.H. 7,6-51, 1M) Bureau of Industrial Hygiene, Connecticut State Department of Health.
27. _____, *Industrial Dermatoses*, American Society of Lubrication Engineers, 1957.
28. _____, "Good Plant Practice for Workers Using Petroleum Products," *American Petroleum Institute Publication 1531* (February, 1959). Sponsored by the Industrial Hygiene Foundation of America.
29. M.M. Key, M.D., *et al.*, "Cutting and Grinding Fluids and Their Effects on the Skin," *American Industrial Hygiene Association Journal* (September-October, 1966), pp. 423-427.

Chapter 6
QUALITY CONTROL OF METAL CUTTING FLUIDS
(WASTE MINIMIZATION AND MANAGEMENT)

In order for a fluid to function as the manufacturer claims it will, the user must observe certain basic precautions. These include care during in-plant handling, storage, machine cleaning procedures, bacterial and water control, fluid recovery and reconditioning, and disposal of used fluids. As the user exercises quality control in his product, he must also take the proper steps to ensure that his fluid is giving him optimum performance. The responsibility for such performance lies with both the manufacturer and the user as described in the following sections.

HANDLING METHODS

Storage

The quality control of metal cutting fluids begins with good storage conditions, and indoor storage of drummed material is best. The extremes of temperature should be avoided; the ideal storage temperature being between 50° and 120° F (10° and 48.9°C). If outdoor storage is unavoidable, the drums should be covered or set on their sides to avoid water accumulation around the bung. The expansion and contraction of the fluid due to temperature changes can result in water seepage into the drum through the bung hole.

Water miscible fluids require more care in storage than straight oil types. Because of their usually more complex chemical composition, emulsifiable or soluble fluids are more prone to physical changes resulting from low or high temperature storage. Nonetheless, straight oils should be protected from freezing temperatures. In this respect, the Cold Test or pour point is not always a reliable guide to safe storage. A straight cutting oil with a zero or -20° F (-28.9°C) pour point may separate a part of its fatty content on long storage at 30° to 45°F (1.1°to 7.2° C).

Bulk storage, particularly of large quantities, is not always possible indoors or underground. Tanks for outdoor storage can be protected by reflective paints which will minimize temperature variation of the contents.

A means of heating or warming the fluid should be provided to facilitate pumping in cold weather, but overheating must be avoided. Low pressure steam is an excellent heating method, but connections must be tight to avoid water contamination due to leakage.

The quality of emulsion can be changed by the attempted emulsification of "soluble oils" after being stored at cold temperatures for long periods of time. Chemical types of water miscible fluids may stratify or separate at low temperatures.

Where large quantities of fluids are consumed, bulk storage is unquestionably best. Bulk fluids require less handling, less storage space, and present no drum return problems. In addition, the cost of fluid in bulk is lower. The quantity of water miscible fluids should be selected to provide for no more than three months total storage time and it is good practice to use two smaller tanks rather than a single large one. By this means, it is easier to change from one type of fluid to another, should the need arise.

Central Versus Individual Systems

The decision to employ a central system depends first on the fluid requirements. A central system simplifies concentration control and cleansing but may result in a compromise fluid or a compromise concentration, or both. Central systems are best suited to mass production facilities where the number of operations and metals machined may be grouped for efficient use of a common fluid and fluid concentration. Where these requirements are met, the higher cost of the initial installation of a central system is more than offset by lower operating costs.

WATER SOURCES AND COMPOSITION

The quality of water is of extreme importance to the efficient use of aqueous metal cutting fluids. The life of the system, filter efficiency, foam characteristics, and even tool life and finish, are influenced by the quality of the water. It is essential that the quality of the available water be studied and the proper water miscible coolant be selected on the basis of the local water characteristics.

Water used for making coolant mixtures should be made as pure as possible for the most economical and trouble-free use, but even the "cleanest" of shop water is rarely pure. Water throughout the United States is usually contaminated with "hardness" minerals or salts, or both, which have a detrimental effect upon the cutting and grinding fluid mixtures made from them. Rain water is soft, containing no minerals. Water obtained from lakes and rivers may be relatively free of minerals or be heavily contaminated, depending upon whether or not the waters have been able to dissolve minerals during their natural course. Water from wells may be relatively free of contamination but, *usually*, most well water is heavily contaminated with minerals.

Minerals in coolant water can cause corrosion of machine tools and machined parts, can aggravate deposition of residues on machine tools, and can increase the rate at which bacteria and fungi grow in the coolant. The "hardness" of water is calculated on the basis of 17 parts per million of calcium carbonate per U.S. gallon as being equal to one "grain." Hardness is caused almost entirely by calcium and magnesium ions. Other elements, such as iron and aluminum, are minor sources but can produce undesirable corrosion effects far out of proportion to their concentration. Occasionally, hardness can be increased by zinc which has been dissolved from newly galvanized pipes.

Water hardness "uses up" coolant concentrate and tends to force it out of solution. The net effect is that part of the concentrate does not contribute to cutting efficiency; instead, it may appear as a gummy deposit or residue on the machine and parts. In addition, the lost concentrate can cause parts and machines to rust.

Minerals other than hardness "salts," such as chlorides and sulfates, contribute to corrosion or rust and, the higher their concentration, the more of the cutting fluid concentrate is required to prevent corrosion. The sulfates are particularly detrimental because they promote the growth of sulfate-reducing bacteria *Desulfovibrio desulfuricans* (1) which produce a "rotten egg" odor.

A machine coolant sump acts as a "still"; the more the fluid is aerated, the more the water evaporates. As this occurs, the minerals in the water increase, causing more residues to form and corrosion to increase. Usually, fluid "makeup," or additions to the machine sump, are on the order of 5 to 20% per day, depending upon the sump capacity and the severity of the operation. Hence, over the period of a month, solids buildup in the fluid mixture can be three to four times that of the original water. The following are examples of how makeup continually concentrates the solids. A soft water of 3 grains hardness can have a hardness of 12 to 14 grains at the end of one month, and 24 to 27 grains in two months of use. A water of 12 grain hardness would yield water of 48 to 52 grains in one month, or 96 to 104 grains in two months. Therefore, the purer the water for making coolant mixtures is initially, the longer the fluid can be used before gumming and corrosion problems occur.

A simple test to check for water residue involves filling a .050 in. by 2 in. (13 by 51 mm) petri dish with the water, placing it into an oven at 220° F (104.4°C), and drying to observe the residue. This procedure can be repeated four times with another petri dish (replenishing the water after each evaporation) to ascertain the effects of coolant water replacement on a monthly basis. Figure 6-1 shows the results of such drying tests on raw water, zeolite treated ("soft") water, and deionized water, each dried once, another four times (refilled after each evaporation).

Figure 6-1a. *Results of drying test used to check for raw water residue. Specimen on left dried once; specimen on right dried four times (with water replenished after each evaporation). (Courtesy, Master Chemical Corporation)*

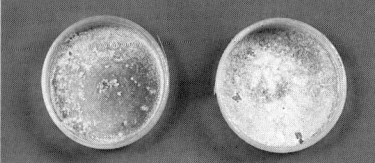

Figure 6-1b. *Results of drying test to check for zeolite treated water residue. Specimen on left dried once; specimen on right dried four times (with water replenished after each evaporation). (Courtesy, Master Chemical Corporation)*

Figure 6-1c. *Results of drying test to check for deionized water residue. Specimen on left dried once; specimen on right dried four times (with water replenished after each evaporation). (Courtesy, Master Chemical Corporation)*

To demonstrate how the continual evaporation of water from a coolant will affect the resulting mixture, see Figures 6-2, 6-3, and 6-4. Figure 6-2 shows a "soluble oil" in distilled water *(a)*, and in a hard water (8 grains) *(b)*. Figure 6-2 *(c)* is the hard water *(b)* evaporated four times, and eight times *(d)*. This photo shows the "soluble oil" which was simply poured on the surface of the respective waters and allowed to stand for several hours. Figure 6-3 shows the samples shaken thoroughly and photographed immediately. Fig. 6-4 shows the thoroughly mixed samples after standing overnight.

These photographs illustrate that minerals (hardness and/or salts) are very detrimental to the stability of coolant mixtures. The more concentrated these minerals are to begin with, the faster they affect the fluids adversely and the more rapidly they build up to cause instability. In fact, minerals can become so troublesome that the concentrate will not *mix* properly, or that coolant tanks would have to be dumped and refilled every week to prevent gumming and corrosion problems.

Minerals in water not only cause residues to form and corrosion to occur, but they also help bacteria to grow. These aspects are among the most important considerations in water miscible fluid usage, and the results can be of substantial economic effect. One method of removing "hardness" from water is to run it through a zeolite softener. However, this process replaces the "hardness" with ordinary monovalent salt so that residues can build up as before, and corrosion can be even more of a problem.

Another method by which minerals are removed from water is reverse osmosis. Reverse osmosis (or R.O.) is a process that uses relatively high pressure (typically 200-800 psi or 1.3 to 5.5 MPa) to force water, containing dissolved minerals and suspended solids, through a semipermeable membrane to produce a high purity water.

Figure 6-2. *Demonstration of how continual evaporation of water from coolant mixture affects resulting mixture. Jar (a) is "soluble oil" in distilled water; (b) in a hard Michigan water; (c) is (b) evaporated four times; and (d) is (b) evaporated eight times. (Courtesy, Master Chemical Corporation)*

Figure 6-3. *Sample as Figure 6-2 but shaken thoroughly and photographed immediately. (Courtesy, Master Chemical Corporation)*

Figure 6-4. *Same samples as Figure 6-3 after standing overnight. (Courtesy, Master Chemical Corporation)*

Pure water can also be produced by deionization (2), which removes all minerals by chemical absorption so that the effluent is equivalent to distilled water. With this method, no residues are left by evaporation of the water and corrosion effects from minerals are eliminated.

Recommended techniques are the use of reverse osmosis or deionization. Many shops which have plating operations already have deionizer installations so that this water source can easily be used for coolant mixes. If not, the cost of installing a satisfactory unit, which will produce trouble-free coolants, is more economical than the cost of rusted machines and workpieces.

Some adjustments can be made in formulating a product to resist the effects of bad water for a period of time. With mineral-free water, however, neglecting other factors, the coolants could be run indefinitely. Figure 6-5 shows the hard 8 grain water of Figure 6-2 (a) and the same water with the same "soluble oil" adjusted to stay perfectly mixed with that water (b). Therefore, in bad water areas (water over 8 grains hardness as shown in Figure 6-6) consult your supplier for recommendations.

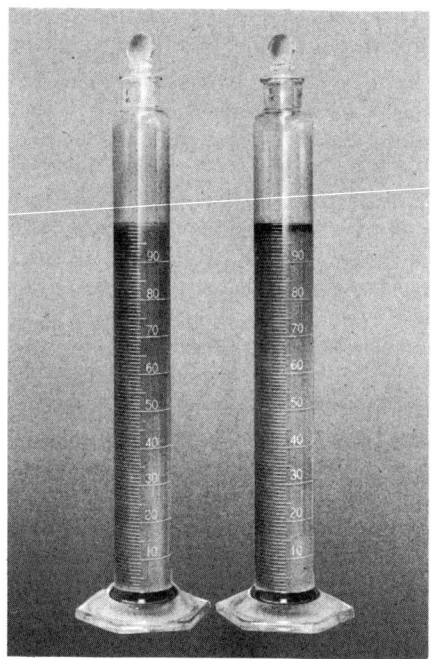

Figure 6-5 *Concentrate mixed with hard Michigan water separates (b). Same "soluble oil" adjusted to stay perfectly mixed (a). (Courtesy, Master Chemical Corporation)*

PROPER MIXING PROCEDURES

Proper mixing procedures are critical to the attainment of long coolant life and economical use of coolant concentrate, as well as to the elimination of coolant concentration related problems. Premixing the coolant concentrate with pure water in accordance with the coolant manufacturer's recommendation assures efficient use of the concentrate.

One type of coolant mixer is the Venturi type. It has the advantage of being inexpensive. However, the coolant water concentration it produces

Water Supply Map & Data

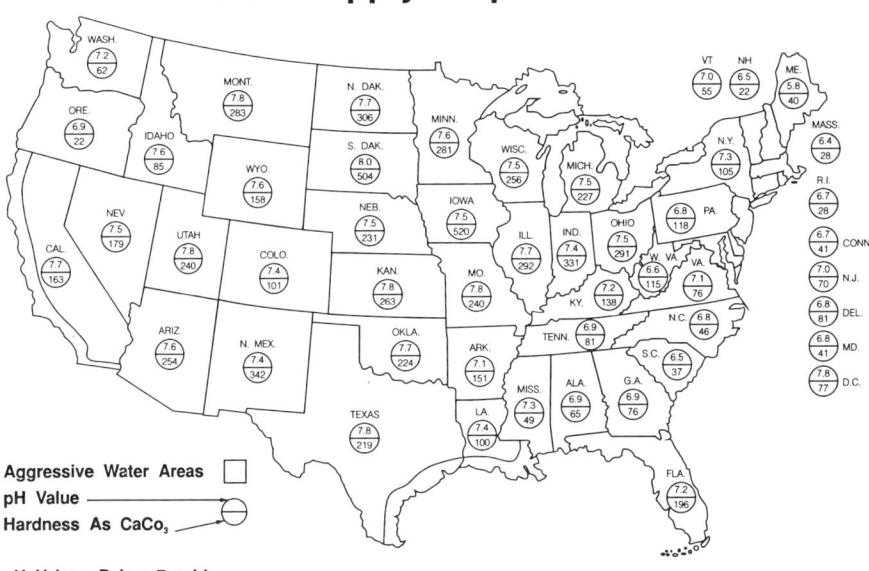

Aggressive Water Areas ☐
pH Value ⎯⎯⎯⎯⎯⎯⎯⎯⎯○
Hardness As CaCo₃ ⎯⎯⎯⎯⎯

pH Values Below 7 acid
        7 neutral
        Above 7 alkaline

Reprinted from **Water Processing for Home, Farm and Business**, written by Wes McGowan, published by and available through Water Quality Association, 4151 Naperville Road, Lisle, IL 60532.

Figure 6-6. *Water hardness by state.*

can fluctuate with changes in the inlet water pressure, coolant concentrate level in the drum, and fluid temperatures.

Another type of coolant mixer is the positive proportioning pump equipped with interconnected water and concentrate cylinders. The coolant concentrate produced with this mixer is not affected by the aforementioned variables.

BACTERIAL EFFECTS AND PREVENTION

Bacteria are of great importance in connection with water miscible fluid usage because bacteria can create a number of problems. Bacteria usually need water in order to grow and they enter the fluid from: (1) water used for mixing, (2) parts, (3) air, (4) operator's hands, and (5) sludge deposits in the coolant sump.

The majority of bacteria need oxygen for growth, and some types multiply by dividing in two approximately every 20 to 30 minutes. Hence, starting with one bacterium and calculating the numbers resulting from its splitting each 20 minutes, the population which would result in 12 hours (provided none would die during that time) is derived as follows:

1 hour 8	10 hours 268,000,000
3 hours 512	11 hours 516,000,000
6 hours 262,000	12 hours 1,032,000,000
9 hours 134,000,000	

While *most* bacteria do not survive in the typical water miscible cutting and grinding fluids, the more important of the less than a dozen bacteria which do, are called:

1. *Escherichia coli*[2]
2. *Klebsiella pneumoniae*[2]
3. *Paracolabactrum species*[2]
4. *Proteus vulgaris*[2]
5. *Pseudomonas aeruginosa*[3]
6. *Pseudomonas oleovorans*[3]
7. *Salmonella typhosa*[1]
8. *Staphylococcus aureus*[1]

Usually more than one is present at a time. These are known as "facultative" aerobic bacteria. They prefer air (oxygen) for best growth, but can adjust themselves to grow in the absence of oxygen.

There is another kind of bacteria, known as anaerobic *(Desulfovibrio desulfuricans)*, which grow in the absence of oxygen, *but do not die* in the presence of oxygen. These bacteria multiply by dividing in two every four hours. They grow much more slowly than the aerobic, but the results of their growth can be very objectionable. They do not grow in fresh clean fluid, but will usually grow once the fluid has been attacked by the aerobic bacteria. In fact, when makeup is added to a "stinking" coolant, it will "freshen up" the mix for a few hours until the aerobic bacteria can break it down.

Lubricating and anticorrosion properties are built into cutting fluid concentrates. When they are properly mixed with water, they will perform their functions of lubricating and cooling and leave an anticorrosion film. However, if bacteria feed on the concentrate in the presence of the water, the lubricating effect is reduced so that cutting tool life and grinding wheel life are gradually reduced, and corrosion can be expected to increase. The reason for this is that when bacteria multiply, they produce acids which cause corrosion. The more rapidly the bacteria multiply, the faster they will attack the fluid. Therefore, the *rate* of bacterial growth is important. If fewer bacteria are grown in a given length of time, harmful effects can be minimized.

The two most troublesome aerobic bacteria are *Pseudomonas oleovorans* and *Pseudomonas aeruginosa*. The bacterium *Pseudomonas oleovorans* lives on oil, so it multiplies rapidly in those machines which leak lubricating and hydraulic oils. Consequently, all steps should be taken to reduce such oil leakage. If leakage cannot be prevented, the free floating tramp oils should

---

[1] Limited growth in a few products.

[2] Grows abundantly in many products.

[3] Grows abundantly in all products

be skimmed from the surface, and the partially emulsified and mechanically mixed tramp oil should be removed by centrifugation.

The bacterium *Pseudomonas aeruginosa* lives on practically anything: minerals in the water, coolant concentrate, discarded food, oils, etc. Note that *Pseudomonas oleovorans* and *Pseudomonas aeruginosa* are both aerobic and facultative and are two species which are present in all water miscible fluids in use because they are difficult to kill.

The main type of anaerobic bacteria which grows in water miscible fluids is *Desulfovibrio desulfuricans*. It is very difficult to control and is found in almost all coolants. It produces a very strong odor of rotten eggs (hydrogen sulfide, commonly known as "Monday morning stink"), and can cause severe dark staining of machines and workpieces. In the presence of iron it can eventually make the fluid turn a black color. When this occurs, it is practically impossible for an operator to stand by the machine because of the foul odors.

One of the most important ways of overcoming the problems produced by *Desulfovibrio desulfuricans* is for the cutting fluid manufacturer to be selective in the raw materials used in the product. Bacteria are very specific in their action and some materials are better food for them than others as shown in Figure 6-7. Thus, the raw materials used should be those which are most resistant to bacterial degradation.

Second, the incorporation of effective microbicides is also helpful in preventing or retarding degradation caused by bacterial action. Very few microbicides are effective, however, and they must be used with great care. Some microbicides which function very well in clean products can actually serve as food for the various types of bacteria found in water miscible fluids which are so easily contaminated.

Third, it is extremely important to maintain *good housekeeping*; in fact, this is the most important way to control bacterial growth. It does little good to put fresh coolant into a dirty machine. All this accomplishes is to give fresh food to the bacteria which are in the sump, on machine surfaces, and in coolant circulating systems. If the machine is thoroughly cleaned with an appropriate cleaner, thoroughly rinsed, and filled with clean, fresh fluid, fluid life will be greatly increased.

Central Systems

On a single machine of approximately 25 gallons (95 liters) capacity, usually from one to five gallons makeup per day is added, based on an average of 5 to 20% makeup per day. Although central systems will generally require three to five times the amount of fluid used by an individual machine, when multiplied by the number of individual machines, much less fresh makeup fluid is required to maintain fluid concentration in a central system. The result is that, if the bacteria in a central system multiply at the usual rate, the smaller ratio of fresh makeup

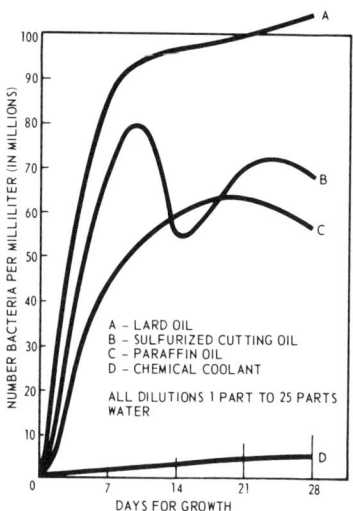

Figure 6-7. *Showing the relationship between food source (raw material) and rate of bacterial growth. (Courtesy, Master Chemical Corporation)*

will not act as effectively as a microbicide against them; hence the net growth will result in more numbers in a given volume of coolant mix compared with the sump of individual machines.

Further, bacteria tend to settle to the bottom of tanks. Therefore, in any system where the fine metal particles and other silt settles, so do the bacteria. This combination helps the anaerobic bacteria to grow best, i.e., where it is farthest from the air (oxygen). Consequently, it is usually more difficult to control bacterial growth in large central systems than in single machine sumps. However, proper cleaning of central system sumps between coolant charges; effective filtration; continuous removal of tramp oils by centrifugation; properly designed, clean coolant headers (preventing sludge buildup); effective flushing of return flumes; and the judicious use of the proper microbicide can keep bacterial growth under control.

There are a number of laboratory methods used to check for potential growth of bacteria in metalworking fluids. Usually, the more concentrated the mixture, the less tendency there is for bacteria to grow. This is indicated by Figure 6-8 showing 4%, 2%, and 1% solutions of concentrate in water. As shown, the more the bacteria, the more discoloration and the more gas produced.

Figure 6-8. *Solutions of 4%, 2%, and 1% concentrate in water with resulting bacterial growth, discoloration, and gas product. The more concentrated the mixture, the less tendency toward bacterial growth. (Courtesy, Master Chemical Corporation)*

FUNGI

Fungi contamination can also present a problem. In general, there is a natural antagonism between various bacteria and fungi. Therefore, it is necessary that, in keeping bacteria under control, fungi are not permitted to flourish. It is a well known fact that the uncontrolled use of antibiotics to kill bacteria in people will result in fungi taking over and creating problems equal to, if not greater than, the original infection.

Minerals in the water help to feed fungi just as they do bacteria, so that removal of minerals from the water also helps to keep fungi under control.

FLUID CLEANING METHODS

A clean cutting fluid system is essential to efficient operations. The means necessary to maintain clean fluid depend on the operation, the metal, and the fluid. Straining is elemental and essential in all operations as a first step. Cleansing by stratification, i.e., settling or skimming, is only efficient for large gravity difference contaminants. Its efficiency is improved by, or requires, relatively large volumes of fluid or long retention times. This is adequate in many instances. If the need for removal of finer particles or micro debris is involved, then more sophisticated means of filtration, centrifuging, and magnetic separation become important.

The cleaning or maintenance of a cutting fluid differs, depending on whether it is straight oil or an aqueous solution. The metal being machined and the operation are equally important considerations. The operation and the metal will indicate the type and fineness of the debris. Ductile metals produce coarse debris that is efficiently removed by straining or settling. Brittle metals or finishing operations produce fine debris which may require filtration.

In addition to the type of operation and the metal machined, the characteristics of the straight oil cutting fluid will dictate the need for treatment. Highly fortified fluids can generate acidic constituents. An excessive accumulation will require absorption filtration to permit continued satisfactory use.

Water may also contaminate oil-type fluids causing corrosion. It can be removed by a water-selective media filter or by high speed centrifugation. Removal of water and acidic decomposition products is generally accomplished by batch-type operations.

Where fine debris results from the metal or the operation, continuous full flow or bypass filtration will be required. Sedimentation or straining are only satisfactory for very coarse particles.

The cleaning and maintenance of aqueous systems generally require more thought, more planning, and greater elaboration than oil-type systems. Where, with straight oil fluids, viscosity is the essential characteristic to be considered in any method of cleansing, many factors are involved in maintaining water systems.

Straining

All fluids, oil or water, require straining. This is essential to pump protection. Double strainers should be inserted and kept free of clogging elements, such as rags or lint. With aqueous systems, stainless mesh strainers should be used.

Settling

Particle size and retention time are the two most important considerations in settling. This method of cleaning and maintenance is generally limited to large sumps or central systems. It is often an essential part of other cleansing methods to reduce the burden on filters, centrifuges, or magnetic separators.

The efficiency of gravity separators is markedly improved by baffles, both above and below the surface of the fluid level. Tramp oils, scums, and hard water soaps may be surface-collected by a baffle and subsequently skimmed. The skimming may be continuous or intermittent. Intermittent removal of floating accumulations may be effected by increasing the fluid level and the use of scum gutters. Continuous removal can be accomplished by a surface paddle or sweep. Dense debris or sediment may be removed continuously by a drag chain, or periodically by emptying the system and shoveling out the accumulated sediment.

Solids-From-Liquid Centrifugation

Solids-from-liquid centrifugation (clarification) is an accelerated settling process and is largely limited to low solids content removal. It may be used to enhance the efficiency of low volume systems or systems too small for efficient settling. It will remove particles too fine for removal at existing flow rates. Centrifugal cleansing is used most often on bypass and to meet specific needs.

Magnetic Separators

Magnetic separators are very effective for removal of ferrous or other magnetic metals. The most widely used is the drum-type with a doctor blade for scraping off the adhering particles. Magnetic separators are most efficiently used with low viscosity fluids or aqueous systems. Where the magnetic metal particles are coated with oil from an emulsion, there is some adherence of nonmagnetic debris. Depending on size and cutting fluid flow rate, the magnetic separator may be used full flow or on bypass. It may be a part of a central system in-line for individual machines. In special instances, it may serve as an important adjunct to other cleaning methods.

Filtration

Where fine straining ends and coarse filtration begins is a moot question. A porous fabric is fundamental to industrial filtration (as opposed to bed-type). The pore size or opening of the fabric will determine the particle size which may be removed. The fabric may be made from woven fiber or extremely fine wire. Or, it may be matted material such as paper or felt.

For any pore size of the filter, any coating of oil on the particles to be filtered can have a decided effect upon drainage of water from the filtered particles or swarf. Hence, leakage of hydraulic or lubricating oils into the coolant system can alter the filtering rate considerably.

The most common filtering systems, where large volumes of fluids are involved, consist of self-advancing rolled fabric. They are actuated by pressure- or vacuum-sensitive devices. This type of area filtration is far more prevalent than fixed-plate or tubular filters. Figure 6-9 shows a vacuum filter unit.

Filtration may be enhanced in terms of through-put by vacuum or negative pressure. The effectiveness of filtration in terms of minimum particle size removal can be improved by supplemental coatings or filter media. Media such as fuller's or diatomaceous earth, highly surface-active silica particles, and others, add depth or apply the bed principle to area filtration. Filter media are rarely used except with fixed plate or tubular

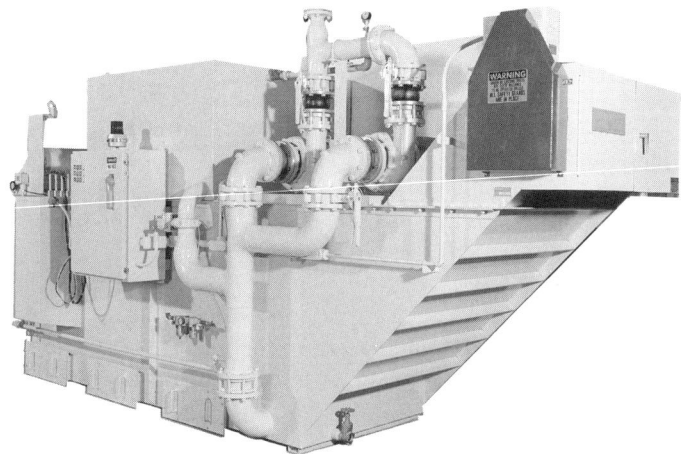

Figure 6-9. *Hyrdroflow vacuum filter unit. (Courtesy, Hydroflow Incorporated)*

areas.

Nonpermeable filtration, sometimes termed "edge filtration," involves discs or stacked plates; thus, it is filtration by clearance. Clearance filtration is readily made self-cleaning by scraper blades. To some extent, this may be regarded as ultrafine straining.

CLEANING MACHINE TOOLS AND CIRCULATING SYSTEMS

The need for cleaning machine tools can be highly dependent on the efficiency of the cutting fluid cleaning process. As improved methods are employed, there is less need for total cleansing of the machine tool or circulating system.

In most instances, systems using straight oils require only periodic de-gumming by means of solvent or detergent additives. Fine particles or chips require occasional removal from the sumps of individual machines.

Water diluted concentrates present an entirely different situation and a need for more frequent and more thorough cleansing of the central system. In aqueous systems, bacterial buildup on machines, the walls of sumps (especially concrete sumps), and circulating systems, requires proper cleaning at regulated intervals. This should be planned on the basis of experience. The period may be every three months for some systems, or yearly for others. Central systems are self-cleansing to a variable extent. Where individual machines are involved, sump cleaning frequency will depend on many criteria, such as sump size, sump design, and amount of contamination. Nonetheless, individual machine tools will require cleansing apart from the central system as a whole.

The two aspects of cleaning machine tool systems are physical and biological. The physical cleaning entails removal of metallic fines, oxidized oil, and other such sludge-forming matter. For this purpose, solvent emulsion cleaners are sometimes effective, but products which combine detergency and solvency are much more thorough. For combined physical and biological cleansing, an antiseptic agent may be added to the cleaning mixture or a machine cleaner, containing detergents and microbicides in an effective concentration, may be selected.

Steam alone, or in combination with chemicals, is sometimes used for cleaning exterior surfaces and is quite effective. Due to the inaccessibility of parts of machine interiors, steam cleaning is not effective for cleaning interior recesses. A properly designed sump, which provides good accessibility for cleaning while minimizing the likelihood of sludge buildup, is extremely important to the success of the cleaning process.

DISPOSAL OF WATER-BASE FLUIDS

Straight oil products properly selected and maintained should not require disposal. Rather than disposal, batch-type recovery should be the rule. Filtration by properly selected media and/or centrifugation and regeneration by the addition of base concentrates usually restores adequate quality for continued use.

Aqueous systems disposal has increasingly become a challenge to industry. Many state, federal, and local regulations prohibit discharge of used fluids without extensive treatment. These regulations may include, but are not limited to, maximum allowable levels for BOD, COD, heavy metals, suspended solids, phosphorus, chloride, oil and grease. Consult with your coolant supplier and the local waste treatment authorities to determine proper treatment methods.

WASTE MINIMIZATION

Today's environment and economic considerations make a coolant management program that integrates the various aspects of this chapter essential for any responsible metalworking manufacturer. Laws are becoming more stringent, haul-away costs continue to increase, and liability remains with the waste generator. These considerations can be addressed at three points.

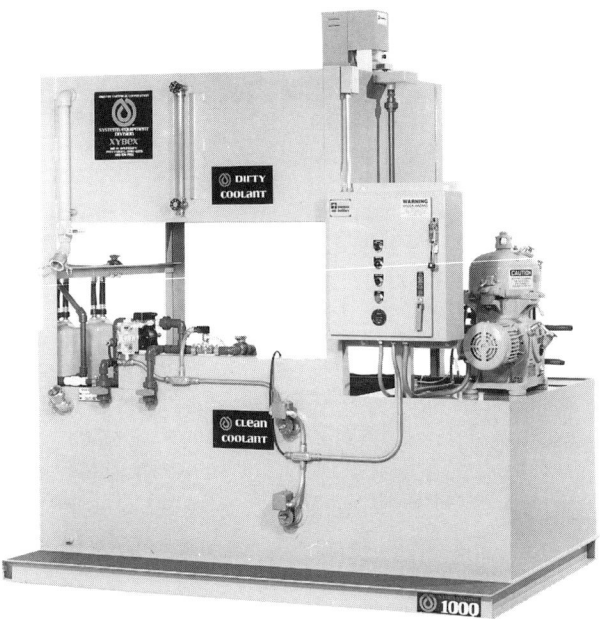

Figure 6-10. *XYBEX® 1000 recycling unit with centrifuge and oil skimmer. (Courtesy, Master Chemical Corporation)*

The first opportunity for implementation of a successful coolant management program is during the design phase of any new system or facility. Properly designed machine tool sumps, coolant headers and flumes will help minimize the presence of coolant failure mechanisms.

The second opportunity to minimize the waste coolant generated is accomplished by proper filtration, as well as properly designed and sized coolant recycling systems that address the problems mentioned in this chapter. One such system is seen in Figure 6-10.

Finally, having implemented the aforementioned steps, the amount of waste coolant produced will be drastically reduced. However, the waste coolant that is produced must be treated and disposed of in an efficient manner that complies with all applicable regulations.

REFERENCES

1. H. W. Rossmore, *et al.*, "Growth Studies on *Desulfovibrio desulfuricans*," *Developments in Industrial Microbiology*, 5 (1964), American Institute of Biological Sciences, Washington, D.C.
2. Joseph Thompson, "Ion Exchange Water Conditioning Methods," (New York: Illinois Water Treatment Company).

CHAPTER 7
COMPATIBILITY OF LUBRICANTS WITH CUTTING/GRINDING FLUIDS

Contamination of cutting and grinding fluids by machine tool lubricants and vice versa can be a serious problem. If the problem of coolant contamination of lube oils is ignored, costly repair of machines will result, together with increased machine downtime and loss of production. Additionally, contamination of coolants by lubricating and hydraulic oils will shorten coolant life by promoting bacterial growth and reducing cooling properties also leading to increased downtime and loss of production. Exact rules cannot be given as to when an oil should be removed because this depends on the nature and the amount of contamination. In general, however, either the machine lubricant or the cutting and grinding fluid should be removed when the important characteristics of either fluid have been seriously degraded. The use of effective filters, the replacement of leaking seals, and in some cases the use of multipurpose oils will lessen or relieve the problem of contamination. When replacing the fluid from a contaminated system, it is a good practice to clean and flush the system before recharging with new fluid.

LUBRICANT SYSTEMS

Machine designers have made great strides in developing machine tools that are capable of turning out finished products with amazing speed and precision. One very important factor contributing to the performance of modern machine tools is effective lubrication of both the machine tool and the part being machined. Lubricants are available which have been formulated to have specific properties in order to meet the demands placed on them. The lubrication requirements of the machine are usually quite different from those required to lubricate the part being machined. Unfortunately, in spite of maintaining separate lubricating and coolant systems, leakage of the lubricants from one system to the other does occur.

Because of the large number and variety of machine tools it is impossible to discuss the lubricants used for each individual machine and the compatibility of these lubricants with cutting and grinding fluids in a single chapter. Fortunately, there are a number of basic systems common to almost all machine tools. Essentially these are hydraulic systems, spindles and bearings, gears, and ways. The significant properties of the lubricant used in these systems, together with their possible interrelated effects with cutting and grinding fluids, are discussed in the following sections.

Hydraulic Systems

Straight mineral oils and, more often, compounded oils are generally used in hydraulic systems. In some instances, *where the danger exists of a fire occurring should a hydraulic line rupture,* a fire-resistant hydraulic fluid is desirable. Practically all hydraulic pumps used on machine tools require the use of two viscosities of oil: either 150 or 330 Saybolt Universal Seconds at 100°F (37.8°C).

Quality hydraulic oils contain additives to supplement and improve certain of their important properties. These are oxidation stability, corrosion inhibition, demulsibility or the ability to separate freely from water, and foam resistance. In addition, the oil must lubricate moving parts of the system so as to reduce wear to an absolute minimum.

In recent years increased emphasis has been placed on the antiwear performance expected from hydraulic lubricants. While conventional hydraulic oils have continued to be completely satisfactory for the majority of hydraulic applications, it appeared necessary to use lubricants containing special antiwear additives in certain equipment operating at relatively high pressure and/or speed. Unfortunately, these special antiwear additives, usually zinc dithiophosphate or (ZDP), promote bacterial growth in coolants and result in high contamination rates.

Heavy duty motor oils were found to give excellent performance in many instances; however, they do not separate from water easily, a characteristic considered objectionable by some users. This may not cause any particular problem in those systems where water contamination is low or in small systems where the oil is frequently drained and replaced. As a general rule, however, rapid separation of water from oils used in hydraulic service is a highly desirable characteristic.

Perhaps the most important property of a hydraulic oil (from a quality standpoint) is its oxidation resistance. This property may be regarded as the lifeline of the fluid, since oxidation is the greatest factor that shortens the life of the oil.

Contaminants such as cutting and grinding fluids, dirt, water, paint, joint compounds, and insoluble oxidation products may all accelerate the rate of oil oxidation. Metals, particularly copper, are known catalysts for oil oxidation and their catalytic effect is greatly enhanced by the presence of water. Active sulfur from cutting oils can corrode copper lines and, in the presence of water, can corrode other system metals as well. Soaps from soluble oils and fatty oils degrade the demulsibility and foam resistance characteristics of hydraulic oils. Dirt or abrasive material in any form is harmful to the closely fitted parts of pumps and controls. It may also lodge in valves and lines, slowing down the flow of the hydraulic fluids and preventing proper functioning of the system.

Fire-resistant hydraulic fluids (commonly used in forging presses but not in machine tools) may be classified in general as being of two basic types. The first type is the aqueous-base fluid which includes water-in-oil (invert) emulsions and the water/glycol mixtures. The second type is the non-aqueous and includes phosphate esters, halogenated hydrocarbons, mixtures of the two, and silicones.

Contamination of emulsion type fluids by cutting coolants can cause separation of the emulsion, resulting in loss of lubrication and rust inhibition qualities. In systems using the non-aqueous type, cutting oils may have an adverse effect on the seal material.

Leakage of hydraulic oil into the cutting or grinding fluid is usually encountered more frequently than the reverse. The primary effect, if oils are employed, is one of dilution. The percentage of active ingredients is reduced and the fluids lose some of their effectiveness. This, in turn, results in poor tool life and finish. When soluble oil emulsions are used as the coolants, instability, odor, short emulsion life, and filtering problems may result. If the emulsions are used for grinding, wheel loading may occur.

Spindles and Bearings

Spindles and bearings may be either grease- or oil-lubricated, depending upon the design. Oil-lubricated spindles require products ranging in viscosity from 40 to 300 Saybolt Universal Seconds at 100°F. (37.8°C). In systems where the oil is recirculated, it is essential that the product be of good quality. The oil should resist oxidation, not emulsify with water, quickly free itself of any entrained contaminants such as air and grinding dust, and prevent the formation of rust. Usually "turbine grade" oils have proved satisfactory. Where there is a possibility of oil leakage, or under conditions where dirt, dust, and other contaminants can get into the bearing, grease is often used in a prepacked, shielded bearing.

The interrelated effects of contamination of oils used on spindles and bearings with cutting fluids are essentially the same as those encountered with hydraulic oils. High-speed spindle bearings can be ruined by small particles of dirt. Every effort should be made to keep contaminants out of circulating oils and, if the oil does become contaminated, it should be changed immediately.

Gears

Types of gear lubricants fall within two classifications: straight mineral oil and extreme pressure lubricants. Straight mineral oil gear lubricants should be well refined and highly resistant to oxidation. In addition, they should be so processed that they do not foam when mixed with air nor emulsify when contaminated with water. Extreme pressure gear oils should be noncorrosive to copper or steel, have a high film strength, be unaffected by the presence of water, and should not thicken or oxidize appreciably in service.

Leakage of cutting coolants into gear lubricants could result in the gear lubricant becoming corrosive to system metals, in loss of oxidation resistance, and in the formation of stable emulsions. Dilution of the gear lubricants by the cutting oils, which usually have a much lower viscosity, will reduce film strength and increase wear.

Gear oil leakage into the cutting coolant can cause staining of parts and make cleaning of parts more difficult. Water miscible coolant life can be greatly reduced by contamination of gear oils because of separation and bacterial growth.

Since the 1970s, a number of NC and CNC machines have been introduced whose gear boxes feature electromechanical disk type clutches very similar in design and function to the disk clutches in automobile automatic transmissions. Because of this similarity, many of these gear boxes require the use of highly compounded automotive type automatic transmission fluid (ATF). While excellent for performing its intended functions in machine gear boxes, leakage of ATF into water miscible coolant has disastrous effects on the coolant. It promotes rapid bacterial growth and lowers the pH of the coolant rapidly, causing staining and corrosion problems on workpieces and bare metal parts on machines. Most importantly, in the presence of water, ATF hydralizes to release acids into the coolants leading to severe skin irritation. Leakage of ATF into water miscible coolants must be corrected immediately.

Ways

A variety of lubricants may be used on ways, depending both on the load and the method of application. Where loads do not exceed approximately 10 lbs/sq in., a straight mineral oil having a viscosity of about 300 Saybolt Universal Seconds at 100°F (37.8°C) is used. When loads exceed 10 lbs/sq in., it is usually necessary to use either a high-viscosity straight mineral oil or, more often, a compounded oil. A way lubricant should prevent chatter, should not press out upon standing, and should prevent metal-to-metal contact between localized high spots.

In most instances, contamination of the cutting fluids by the way lubricant should be very small and not much of a problem. Some way lubricants, however, may tend to be washed off by some cutting fluids, resulting in chattering or stick-slip motion of the carriage.

TOTAL LOSS LUBE SYSTEMS

The vast majority of machine tools built since the 1970s feature slide way lubrication from centralized automatic "total loss" lubrication systems. While these do an excellent job of assuring adequate lubrication of slide ways, the relatively large amount of slide way oil they displace ultimately contaminates the coolant, again promoting bacterial growth, reducing cooling ability and increasing fluid disposal. Since it is necessary to assure adequate lubrication of slide ways, new machines frequently have the lubrication systems set to the maximum flow setting for the initial few months of operation. However, after a machine has "worn-in," the lubrication system should be adjusted to the recommended settings in the Operation & Maintenance Manual to minimize slide way oil contamination of the coolant. Way oil should be skimmed off the surface of the coolant by the installation of skimming wheels or belts to increase coolant life.

MULTIPURPOSE MACHINE TOOL LUBRICANTS

Reference has been made to a condition that exists in many machines, i.e., the leakage of the machine lubricant into the cutting fluid system and, to a lesser extent, leakage of the cutting fluid into the lubricating system. Although machine lubricant leakage does not always occur on every machine, it happens more frequently than not, and becomes more frequent as the machine becomes older. Continual dilution of the cutting fluid with a machine lubricant will ultimately reduce the percentage of active ingredients in the cutting fluid to the point where it is no longer effective. On the other hand, dilution of the machine lubricant with some types of cutting fluids can promote sludge formation and corrosion, and eventually ruin machine parts.

To overcome this problem, "dual purpose oils" were developed which could be used both as general machine lubricants and as cutting fluids. Eventually their use was extended to the hydraulic systems, resulting in their accepted classification as "tripurpose oils" or "multipurpose oils."

These multipurpose oils combine the characteristics of good lubricating oils with certain properties essential to cutting fluids. They are noncorrosive to steel or copper, have good resistance to oxidation and foaming, and good water separating characteristics. In addition, they give excellent performance on machining operations which are considered mild or slightly severe. Even on severe operations, where the cutting fluid employed may have to be more highly compounded, the effect of dilution from that portion used in the lubricating system will be far less than that from conventional machine lubricants.

One word of caution regarding the use of multipurpose oils; it is essential that only fresh clean oil be added to the lubricating side of the machine. Oil from the coolant reservoir should never be added to the lubricant reservoir, since the chips and fine metal particles contained therein would cause damage to the bearings, pumps, and other precision parts.

An added advantage obtained when multipurpose oils are used is the reduction of lubricant inventory and simplified lubrication.

Chapter 8
ECONOMIC CONSIDERATIONS IN CHOOSING A FLUID

The physical aspect of cutting and grinding fluids -- chemical, metallurgical, and other technological factors working together -- make a successful metal cutting operation. Although specific physical details may be difficult to extend to a particular problem, the physical study of a production operation must be made on a "total systems" basis. No single physical factor can be changed without realigning the relative importance of all other physical factors that comprise the system. And, within the bounds of making a quality end product, the question of whether a change has "helped or hurt" should be investigated analytically.

The thesis of this chapter is that most physical factors are also *economic* determinants. Economic analysis must be made on a *total systems* basis. Economics can and should be the common denominator for evaluation, once physical limits are ascertained. Whether it is mouse traps or missiles, a good measure of *total manufacturing cost per unit* will reveal if changes are helping or hurting, and keep the operation on a systems basis.

Conventional treatment of cutting and grinding fluid expenditures places them in a position of minor importance in terms of dollars expended in the total manufacturing cost picture. Typically regarded as a maintenance item or a part of direct manufacturing burden expressed and controlled as a percent of direct labor, sorting out cause and effect can be difficult. This is complicated by the fact that central systems feeding various departments and types of operations are usually prorated as to costs borne by each function. Cost reduction efforts are often centered around bulk delivery and handling, and reducing the per-gallon-cost of the fluid.

Chapter 1 pointed out that the key to efficiency in the metal removal process is *what happens where the chip is formed.* Vital to the control of friction, temperatures, built-up edge, tool forces, tool life, surface finish and finished part conformative to specified geometry, fluids are often underrated in their importance. To avoid extending this to the economic systems, a redefinition of the elements of manufacturing costs can be made into two broad categories: (1) costs *associated with* a cutting and grinding fluid, and (2) costs *affected by* a cutting and grinding fluid. Expenditures can then be subdivided into *direct* and *indirect* costs with respect to these categories.

Some liberties are taken in this discussion with the traditional accounting terms of "direct" and "indirect" with respect to cutting fluids. Using this method of classification, elements of total manufacturing costs are shown in Table 8-1.

Some of the relationships between elements outlined in Table 8-1 warrant explanation. For instance, Number 3, "Frequency of dumping, cleaning, and recharging," is a function of type of fluid, chemical stability, susceptibility to bacterial action, type of material being cut, efficiency of filtration method used, mechanical effect of filtration, or external contamination such as hydraulic oil leaks.

The necessity for Number 4, "Maintenance additives," depends on considerations similar to those enumerated under Number 3. Microbicides, pH buffers, emulsifiers, wetting agents, etc., which are not a part of the initial cost of fluid makeup, are in this category.

Number 5, "Control and handling procedures" required by a fluid installation vary widely depending on fluid type and other conditions.

The tool degradation resulting from the generation of the work surface (Number 6, "Perishable tool life") is a generally recognized cost. Tool reconditioning costs must include tool resharpening and grinding wheel dressing which are not widely recognized.

Number 8, "Change in rate of operational cycle," affects production capacity, machine utilization, and the balance and utilization of other cost inputs. It is obvious that a change in rate of operation to the "optimum" cycle time will reduce overall costs. Minimum cost is usually preferable. This concept is the heart of the systems analysis approach and leads to the final conclusion: comparison between fluids should be made at optimum conditions for each fluid (i.e., comparing minimum cost rate of operation to minimum cost rate of operation.)

Aside from making a job "easier" or more efficient at the cutting point, the proper cutting fluid affects general working conditions which, in turn, influence output (Number 10, "Labor environment and efficiency"). Safety hazards, respiratory or skin conditions, and foul odors (and sometimes inherent product odors and appearance) caused by cutting and grinding fluids do not aid cost reduction. A fluid change that corrects such conditions may open the door to completely revised operating procedures in a given area, and can be the sole justification of the change.

Number 11, "Machine and tooling condition and tool life," are affected by changes in the amount and type of lubrication they receive, general cleanliness, and corrosion prevention afforded by the cutting or grinding fluid used. Any of these factors may result in a change in the rate at which machines or tooling depreciate over a period of time, affecting their useful life and their contribution to total manufacturing cost.

ECONOMIC JUSTIFICATION

Using the foregoing definitions, economic justification can be explored from several points of view. In typical manufacturing installations, items defined as *affected* costs are usually many times the magnitude of the *associated* costs. Therefore, a change in associated costs is readily justifiable, if change in affected costs can be observed and documented. Likewise, under both associated and affected costs, items defined as *indirect* may be many times larger than those defined as *direct*. The dangers of comparing only direct costs of the system after a fluid change lie in the fact that the cost proportions will be different for each physical installation.

The important point is that *total* manufacturing cost must be computed, including all items of direct and indirect costs, for both associated and affected costs for each method of operation under study. This insures that no single item will influence a decision out of proportion to its economic importance. Table 8-2 is a suggested work sheet to assist in this calculation.

Once engineering improvements have been justified and installed, it is advisable that cost inputs be correlated with existing Accounting Department figures. With cause and effect correctly identified, routinely-generated accounting performance ratios may serve as a check on the economic behavior of the system.

The engineer is responsible for selecting any of the physical inputs to the system. Though he may be a Lubrication Engineer, Tool Engineer, or Equipment Engineer, he has economic responsibility to the total system, not to any one item or area of interest.

To appreciate the proportional cost picture, a set of typical production cost circumstances of a metal removal operation is assumed for the purpose of examining the economic impact of improvement efforts. All results are tabulated in Table 8-3 for six examples. Example A is explained in detail.

Example A: Present Conditions, Machine Base Supply, Dumped for Dirt

Operating Specifications

A battery of ten machine tools, which cost $50,000 each, operate eight hours per shift, two shifts per day, five days per week, fifty weeks per year. Production is 1000 workpieces per shift.

The ten machines are tended by three operators on each shift. Labor rate for machine operators is $11 per hour (day rate). Pay rates used here do not include fringe benefits. Fringes may increase payroll cost by two to three times base rates. This applies to all persons on the payroll. Maintenance is provided on an "as needed" basis at an hourly rate of $11.

One hundred gallons of cutting fluid is stored in the base of each machine. The cutting fluid, charged into each machine at the start of each week, is mixed at a 20:1 ratio (20 parts water to 1 part concentrate).

During the week, makeup solution will be required at the start of the shift on days two, three, four and five to replace fluid lost due to drag-out and evaporation during the previous shift. Makeup is mixed at a ratio of 25:1 since evaporation results in the proportion of concentrate increasing as parts are produced.

At the end of each workweek, the cutting fluid is pumped out and the machine sumps are cleaned. A total of 950 gallons of waste fluid, 900 gallons of dirty fluid plus 50 gallons of fluid used to clean the systems, are pumped out and hauled away by a waste hauler. The waste hauler charges $0.40 per gallon for the service.

Costs

There are four categories of cost:
1. Associated, Direct Costs
2. Associated, Indirect Costs
3. Affected, Direct Costs
4. Affected, Indirect Costs

Associated, Direct Costs. The direct labor cost per piece, plus the weekly cost of cutting fluid, including the initial charge and fluid makeup, equals the total Associated, Direct Costs.

Direct labor cost per piece. There are three operators per shift, two shifts per day, eight hours per shift, five days per week. The labor rate is $11 per hour for each operator. Total yield is 10,000 workpieces per week. The direct labor cost per piece is:

3 operators x 2 shifts x 8 hours x 5 days x $11 per hour = $2640 labor per week

$2640.00 ÷ 10,000 pieces = $0.264 direct labor cost per piece

Cutting fluid cost per piece. This cost is derived from the cost of the cutting fluid concentrate and the cost of water for mixing. It includes the total cost for the initial charge plus the cost for weekly makeup.

The purchase cost for cutting fluid concentrate is $2 per gallon. Purchases are taxed at 8%, or $0.16 per gallon of concentrate. Cost for inventory and handling is $0.06 per gallon. Total cost assigned to cutting fluid concentrate is:

$2.00 cost + $0.16 tax + $0.06 inventory/handling = $2.22 per gallon

Water is provided from the plant's private well and run through a commercial water deionizer. The cost of water delivered to the mixing station is 0.5 ¢ per gallon.

To fill the sumps for ten machines requires 47.6 gallons of concentrate and 952 gallons of water. Total cost of material for the initial charge is:
   47.6 gallons/concentrate x $2.22 per gallon = $105.67
   952 gallons/water x $0.0025 per gallon = $2.38
   $105.67 + $2.38 = $108.05 cost of material for the initial charge

During the week, makeup solution is required at the start of the shift on days two, three, four and five to replace loss due to evaporation and drag-out during the previous shift. Makeup solution is mixed at a 25:1 ratio. Total cost of material for makeup per week is:
   3.85 gallons/concentrate x $2.22 per gallon = $8.55
   96.15 gallons/water x $0.0025 per gallon = $0.24
   $8.55 + $0.24 = $8.79 cost of makeup per day
   4 days x $8.79 per day = $35.16 makeup per week

The total cost for cutting fluid, including initial charge and four days of makeup, is:
   $108.05 initial charge + $35.16 makeup = $143.21 cutting fluid per week
   $143.21 ÷ 10,000 pieces = $0.0143 per piece

Summary. The total Associated, Direct Cost is:
   $0.0143 cutting fluid + $0.264 direct labor = $0.2783 associated, direct cost per piece

Associated, Indirect Costs. Maintenance charges, for both the system and the machine tools, added to haul-away charges, make up the Associated, Indirect Costs.

System maintenance cost per piece. Maintenance is required to charge the system initially, add the required daily makeup solution, pump out the system, and clean the sumps. These tasks total one hour per machine for a total of 10 hours per week. Maintenance pay rate is $11 per hour. The system maintenance cost per piece is:
   10 hours x $11 per hour = $110 per week
   $110.00 per week ÷ 10,000 pieces = $0.011 system maintenance cost per piece

Machine maintenance cost per piece. The operation uses no additives, has no filtration system (except for a screen used to prevent most chips from passing into the fluid storage of the machine tool), and fluid concentration is not monitored. Maintenance of the machine tool's pumps and piping requires an average of two hours per week at the maintenance rate of $11 per hour. The machine maintenance cost per piece is:
   2 hours x $11 per hour = $22 per week
   $22 per week ÷ 10,000 pieces = $0.0022 machine maintenance cost per piece

Haul-away cost per piece. Used fluid is removed from the premises by a waste hauler who charges $0.40 per gallon for this service. Fluid is pumped out at the end of day five. Normal drag-out and evaporation yield 900 gallons of dirty fluid. Fifty gallons of cleaning fluid are required to clean the sump of each machine in preparation for installing fresh fluid. The

cleaning fluid is added to the dirty fluid, totaling 950 gallons for disposal by the waste hauler. The cost of haul-away:
 950 gallons x $0.40 per gallon = $380.00
 $380.00 ÷ 10,000 pieces = $0.038 haul-away cost per piece

Summary. The total Associated, Indirect Cost is:
 $110 system mtc. + $22 machine mtc. + $380 haul-away = $512
 $512.00 ÷ 10,000 pieces = $0.0512 associated, indirect cost per piece.

NOTE: In all the cost calculations, corrections were not made for the fact that production will be lost when installing fluid, adding makeup, maintaining equipment, and making tool changes. *Be forewarned--these elements have not been included in the calculations, but may contribute significantly to the economic balance.*

Affected, Direct Costs. Tools and tool resharpening make up affected, direct costs. Perishable tools costing $100 are used in this operation. The tools yield an average of 200 pieces before resharpening is necessary. Resharpening cost is $10. The tool can be ground 10 times before it must be discarded (tools are considered to have no scrap value). The tool cost per piece is:
 10 resharpenings x $10 = $100 resharpening cost per tool
 $100 purchase cost + $100 resharpening = $200 total tool cost

Summary. The total Affected, Direct Cost is:
 200 pieces per use x 11 uses = 2200 pieces
 $200 total tool cost ÷ 2200 pieces = $0.091 affected, direct cost per piece

NOTE: The cost per use of a cutting tool usually includes the cost of tool change time. This has not been included in the calculation above. Also, consider the loss in production due to time spent changing tools.

Affected, Indirect Costs. Affected, indirect costs are derived from return on machine tool investment, machine maintenance, inspection tooling replacement, scrap allowance, fixed overhead costs and purchased raw materials.

Return on machine tool investment. Each of the 10 machines in the battery costs $50,000. It is company procedure to set aside 10% of the original cost of investment per year as return on capital invested. This is charged against manufacturing and included in the cost per piece. The 10 machines produce 500,000 pieces per year. The return on machine investment per piece is:
 10 machines x $50,000 = $500,000 machine tool cost
 $500,000 machine tool cost x 10% = $50,000 return on machine investment
 $50,000 ÷ 500,000 pieces = $0.10 return on machine investment per piece

Machine maintenance. Machine maintenance averages one hour per shift at a direct labor cost of $11 per hour. Machine maintenance is:
 $11.00 per hour ÷ 1,000 pieces = $0.011 machine maintenance per piece

ECONOMIC JUSTIFICATION

Other costs. The remaining costs, per piece:
Inspection tooling replacement costs are $0.002.
Allowance for scrap is $0.006.
Fixed manufacturing overhead cost for supervision and quality control are $0.06.
Purchased raw materials are $0.21.

Summary. The total Affected, Indirect Cost is:

Return for machine tool investment	$0.10
Machine maintenance	0.011
Inspection tooling replacement	0.002
Allowance for scrap	0.006
Fixed manufacturing overhead	0.06
Purchased raw materials	0.21
Total affected, indirect cost per piece	$0.389

Summary, Example A. Total cost per piece for the operation in Example A, Present Conditions, Machine Base Supply, Dumped for Dirt is:

Associated, Direct Cost	
Machine operator labor	$0.264
Cutting fluid	0.0143
Total	$0.2783
Associated, Indirect Cost	
System maintenance	$0.011
Haul-away	0.038
Machine maintenance	0.0022
Total	$0.0512
Affected, Direct Cost	
Tool cost	$0.091
Total	$0.091
Affected Indirect Cost	
Return for machine tool investment	$0.10
Machine maintenance	0.011
Inspection tooling replacement	0.002
Allowance for scrap	0.006
Fixed manufacturing overhead	0.06
Purchased raw materials	0.21
Total	$0.389
Total Associated, Direct Cost	$0.2783
Total Associated, Indirect Cost	0.0512
Total Affected, Direct Cost	0.091
Total Affected, Indirect Cost	0.389
Total	$0.8095
Cost per piece for parts produced in Example A	$0.8095

Example B: Central System With Filtration, Same Fluid, Dumped For Rancidity

Operating Specifications

Operations are the same as those in Example A except that a central tank equipped with a simple filter has been installed. The tank provides central storage and is the source for cutting fluid.

Costs

Change in Associated, Direct Costs. In Example A, the cost of cutting fluid concentrate, water, mixing and an initial charge was $108.05. In Example B, cost of each initial charge is still $108.05 but is only incurred once in each four weeks or about 12 times a year ($108.05 x 12 = $1296.60). Cost per piece is $0.0026. Yearly savings are $5402.50 - 1296.00 = $4106.50.

Cost for makeup in Example A was $8.79 per day. ($35.16 per week x 50 weeks = $1758.00). Cost of makeup remains the same in Example B, $0.0039 per piece.

Total cost per piece for cutting fluid is $0.0026 + 0.0036 = $0.0062. Savings in cutting fluid cost are $4105.00. Savings per piece in cutting fluid cost are $0.0081.

Changes in Associated, Indirect Costs. Machine maintenance time is reduced due to the revision in the system. Maintenance no longer pumps out, cleans and recharges individual machine sumps. Pump-out and cleaning is reduced to once every four weeks. Make-up is added at one location rather than 10 which reduces maintenance time to two hours per week rather than the 10 hours per week as in Example A. Total maintenance cost is reduced from $0.011 to $0.0022 per piece for a savings of $0.0088 per piece. Yearly saving is $4400.00.

Cost of haul away is reduced. In Example A, 950 gallons of waste fluid were hauled at a cost of $380 per week or $19,000 per year. With the central tank and filter in Example B, it is only necessary to haul-away 950 gallons every fourth week, for a cost of $4560 per year. The need for 36 haul-aways has been eliminated, saving approximately $14,000 per year. Savings per piece is $0.029.

Changes in Affected, Indirect Costs. The tank and filter cost $8,000. An additional set-aside for return on this investment is $800 per year. This adds $0.0016 to the cost per piece. Costs for set-aside in Example A to cover the machine tool investment of $500,000 was $0.10 per piece. For Example B, $0.0016 is added to this for the tank and filter. Total cost of set-aside per piece for Example B becomes $0.1016.

Summary Example B. Changes in costs that effect the operation in Example B are:

Investment in central tank and filter	$8,000.00
10% set-aside for cost of investment	(800.00)
Added cost per piece due to new investment	0.0016
Reduction in maintenance cost	$4.40
Reduction in haul-away costs	$13,680.00
Reduction in cutting fluid usage	$4,105.00
Net reduction in fluid usage	($21,385.00)
Projected production per week	10,000
Projected production per year	505,000
Total cost per piece for parts produced in Example B	$0.7623

Example C: Same System and Fluid With Close Control of Concentration and pH, Microbicides and Buffers Added

Operating Specifications

The system and cutting fluid remain the same as in Example B, except that here close control on concentration, pH, need for microbicides and buffers is maintained by a chemist. The chemist monitors and makes additions as needed.

The chemist recognizes that the reason for dumping changed from dirt in the fluid (as was the case before the filter was installed), to a situation where rancidity is the problem. The chemist reasons that rancidity can be controlled with proper addition of microbicides.

Costs

Changes in Associated, Indirect Cost. Microbicides costing $10 per week are added. One hour per week of the chemist's time is required. The chemist's pay rate is $15 per hour. Concentration and pH are checked twice a week. Concentrate, water, and/or buffers are added as needed. This requires two additional hours per week.

Costs for chemical control are:
- $10 per week x 4 weeks = $40 microbicide purchase cost
- 1 hour per month x $15 per hour = $15 installation cost
- 8 hours per month x $15 per hour = $120 to check concentration and pH
- $40 + $15 + $120 = $175 total microbicide cost for month
- $175 per month x 12 months = $2100 yearly microbicide cost

Total cost per piece for yearly chemical control is:
- $2100.00 ÷ 505,000 workpieces = $0.0042 cost per piece

All other costs remain the same. Tank life for the fluid increases to 16 weeks. Savings are realized since there are now three fewer pump-outs and haul-aways over the 16 week period. *(Note: Fewer pump-outs improves productivity since downtime is reduced. The value of this is included in the labor cost calculations.)*

Waste hauling charge of $38.00 are incurred each time waste fluids are hauled. Savings on this are 3 x $380 = $1140 in each 16 week period. Counting three (16 week) periods per year projects to a yearly saving of $3420, or $0.0067 per piece produced.

Summary, Example C. The net effect of action in Example C is a further gain but at a cost in the chemist's time. Production is projected to be 510,000 per year at a cost per piece of $0.751.

Example D: Same System With More Stable Fluid Buffers and Microbicides Incorporated in the Concentrate

Operating Specifications

This system is the same as that in Example C except that a more stable cutting fluid which has buffers and micobicides incorporated is used. The cost for this cutting fluid is higher.

The plant chemist concludes there has been insufficient return for the effort expended in Example C. It is believed it would be more efficient to use a cutting fluid that would not require periodic additions other than those needed to maintain concentration control.

Costs

Changes in Associated, Direct Costs. The concentrate supplier is contacted to obtain a medium duty concentrate specifically compounded for stability in a central system. A concentrate in which microbicides and buffers are incorporated is obtained, but at a higher cost. The price per gallon is $4.50. Taxes at 8% are $0.36 per gallon. Total delivered price of concentrate is $4.86 per gallon.

Costs for inventory, handling, mixing and maintaining pumps and piping remain as they were. Demands on the chemist's time are reduced. Fluid consumption stays at the same level as in Example C.

An increase in tool life of about 5% is noted. Decrease in frequency of tool changes results in an increase in output to 1020 pieces per shift. Cost is spread over more pieces, and cost per piece is reduced.

Increased cost of concentrate requires recalculating the cost per piece for fluids charged into the central tank and provided as makeup. This calculation follows the same format as in Example A.

Total cost assigned to cutting fluid concentrate is $4.86 per gallon. Fluid for the initial charge is mixed at a 20:1 ratio; makeup at a 25:1 ratio (these proportions are the same as those used in Examples A,B and C).

To fill the sumps for 10 machines requires 47.6 gallons of concentrate and 952 gallons of water. Total cost for the initial charge is:
 47.6 gallons/concentrate x $4.86 per gallon = $231.34
 952 gallons/water x $0.0025 per gallon = $2.38
 $231.34 + $2.38 = $233.72 cost of material for the initial charge
Total cost of material for makeup per week is:
 3.85 gallons/concentrate x $4.86 per gallon = $18.71
 96.15 gallons/water x $0.0025 per gallon = $0.24
 $18.71 + $0.24 = $18.95 cost of makeup per day.

The total cost for cutting fluid, including initial charge and four days of makeup, for the 16 week period is:
 $18.95 per day x 78 days = $1478.10 cost of makeup
 $233.72 initial charge + $1478.10 makeup = $1711.82 cost for cutting fluid/16 weeks

During the 16 week period, a total of 160 shifts are worked. Output per shift holds at 1020 pieces. Total pieces produced in 16 weeks are 163,000. Cost per piece for cutting fluid is $0.01. This is about 70% of the cost per piece for cutting fluid in Example A. Note the cost per piece for labor also has decreased due to the increase in output per shift.

Summary, Example D. In Example A, total cost per piece was $0.8095. In Example B, the cost per piece improvement can be credited to the use of the central system. Examples C and D show cost savings obtained when improved cutting fluid and fluid control is applied.

Under conditions established in this example production per week is 10,200 and per year is 510,000. Cost per piece is $0.757. Production has increased but cost per piece has changed little.

Example E: Same System, Increased EP Additives To Extend Tool Life

Operating Specifications

The system in Example E is the same as in Example D with the concentrate further enhanced through EP (extreme pressure) additives.

Testing continues with the system as in Example D but with a premium concentrate which has the longevity attributes of the concentrate used in Example D, plus attributes which are intended to improve cutting tool life.

Costs

Changes in Associated, Direct Costs. Concentration and all other conditions remain the same as in the previous example. The concentrate selected costs $9.50 per gallon plus tax of $0.76. Total cost of concentrate is $10.26 per gallon. Note that this concentrate costs about five times that of the original concentrate. Using this concentrate there was no change in attention time, frequency of pump-out or any other condition.

Changes in Other Costs. Cutting tool life improved by nearly 40% over that experienced in Example D. Downtime for tool changes was reduced (longer tool life = fewer tool changes which yields increase in productivity). Output increased to 1050 pieces per shift. The life of the cut-off tool increased dramatically. This is credited to reduction in cutting force with the EP additives. As a result, the width of the cut-off tool was successfully reduced. This move increased the number of pieces yielded per unit length of barstock. Workpiece material cost is reduced by about 2%.

Cutting tool cost is reduced to about $0.055. Fixed manufacturing overhead is spread over a larger number of workpieces as is the labor cost for operators of the machine tools.

Summary, Example E. The action taken in Example E results in a projected cost per piece over a year's time to be $0.647.

Example F: Same System And Fluid, Feed And Speeds Optimized To Minimum Cost Position

Operating Specifications

The same system and cutting fluid as those in Example E are used, but action is planned to take advantage of improvements in cutting tool operation provided by the more expensive but enhanced cutting fluid. The action will be to revise cutting conditions to yield Minimum Cost Production.

Costs

Changes In Costs. Cutting conditions were examined and it was determined that Minimum Cost Production (based on the relationship of tool cost, tool change time, machine tool investment cost, and machine operator pay rate) would be obtained when cutting conditions (cutting speed and cutting feed rate) were adjusted to reduce cutting tool life to one-half the original value. Adjustments were made and output increased to 1365 pieces per shift (an increase of about 30%).

Increasing the rate of metal removal, while within the capacity of the machine tool, was expected to result in more rapid wear of the machine. Projected machine life was reduced to eight years from 10. Machine tool maintenance perhaps should have been, but was not, figured to increase.

Increased output reduced the cost per piece for operator labor, and reduced the cost per piece for fixed manufacturing overhead.

Other elements of cost were not recalculated. For example, even though cutting tools will be run at higher speed and heavier feed tool life, there are 30% more pieces produced per shift without reducing tool life in pieces per grind of the cutting tool.

Summary, Example F. Under conditions established in Example F, output per week becomes 13,650 pieces and per year 682,500 pieces. Cost per piece produced is $0.547.

OVERALL RESULTS

The results of these examples are summarized in Table 8-3 and demonstrate the high leverage effect of proper fluid selection and management on the end cost per unit. This shows that close cooperation with the accounting function is necessary to be sure everyone has an understanding of what comprises *total direct manufacturing costs*. If decisions are made on only *part* of the cost picture, the optimum overall solutions can be missed. The goal is the reduction of *total* manufacturing cost per unit.

Table 8-1. Elements of Total Manufacturing Costs

	Associated Direct	Associated Indirect	Affected Direct	Affected Indirect
1. Cost per gallon	Mfgs. price plus taxes. Also includes in-plant and in-tank handling.			
2. Cost of additions (makeup)	Usage rate (will vary for different fluid types for same job) x cost/gal. + cost of water used.			
3. Frequency of dumping, cleaning, and recharging		Cost/gal. x tank capacity + labor + cost of used fluid disposal.		
4. Maintenance additives		Cost of item + labor		
5. Control and handling procedures		Maintenance or laboratory personnel check + filtration cost + maintenance of tanks, pumps, etc., connected to system.		
6. Perishable tool life (cutting tools, wheels, etc.)			Initial tool cost, resharpening cost, number of resharpenings, tool life expressed in number of parts produced.	
7. Product quality level and consistency				Quality change reflected by increase or decrease in rejects.

ECONOMIC JUSTIFICATION 157

Table 8-1. Elements of Total Manufacturing Costs—Cont'd

8. Change in rate of operation cycle	Systems analysis approach based on concept that increased or decreased performance (e.g., tool life) affects cost relationships with other inputs (e.g., labor cost per part).
9. Raw material utilization	Increases or decreases in cutting efficiency affects stock allowances, raw material recovery rate, and machine cycle times.
10. Labor environment and efficiency	Change of fluid correcting or contributing to unsafe, unhealthful, foul odors, etc., affects efficiency and therefore overall costs.
11. Machine and tooling condition and tool life	Housekeeping practices and choice of fluid affect machine and tooling depreciation factors. Maintenance time and consistent operation of mechanical devices (e.g. loaders and gauges) also influence costs.

Table 8-2. Cutting and Grinding Fluid Justification - Cost Work Sheet

Fluid Type_____   Operation_____

Associated, Direct Costs

- A1. Price/Gal. _____
- +A2. Taxes/Gal. _____
- +A3. Inv. & Hdlg./Gal. _____
- =B1. Total Cost/Gal. _____
- xB2. Gal. Used/Shift _____
- =C. Makeup/Shift($) _____

- D1. Gal. Water/Shift _____
- xD2. Cost of Water/Gal. _____
- =E. Cost of Water/Shift _____

- F. Units Production/Shift _____
- G. Associated Direct Cost/Unit (C+E) ÷ F _____

Associated, Indirect Costs

- H1. System Capacity, Gal. _____
- xH2. Mix Cost/Gal. _____
- =J1. Cost of Initial Charge _____
- +J2. Labor to Charge _____
- +J3. Labor to Clean _____
- +J4. Cleaner Cost _____
- +J5. Cost to Dispose Used Fluid _____
- =K. Cost of System Recharge _____

- L1. Maintenance Additives During Life _____
- +L2. Labor for Additives _____
- +L3. Lab Control Labor/Tank Life _____
- +L4. Mech. Maint. of System/Tank Life _____
- =M. Control & Maint. Cost/Tank Life _____

- N. Tank Life _____
- P. Units Production/Tank life _____

- Q. Associated, Indirect Cost/Unit (K+M) ÷ P _____

Affected, Direct Costs

- R1. Initial Tool Cost _____
- R2. Reconditioning Cost Resharpening Diamond Dressing _____
- R3. No. of Recond./Tool _____
- S. Total Tool Costs R1 + (R2 x R3) _____
- T1. Units Production/Recond. _____
- T2. Units Production/Tool _____

- U. Affected Direct Cost/Unit (S ÷ T2) _____

Affected, Indirect Costs

- V1. Raw Material Cost/Unit _____
- V2. Direct Labor Cost/Unit _____
- V3. Machine Cost/Unit (Depreciation+Interest) _____
- V4. Machine Maintenance/Unit _____
- V5. Size Tooling Cost/Unit _____
- V6. Scrap Cost/Unit _____
- V7. Rework Cost/Unit _____

- W. Affected Indirect Cost/Unit _____
  (V1 + V2 + V3 + V4 + V5 + V6 + V7)

Summary

	Associated Costs	Affected Costs
Direct	(G) _____	(U) _____
Indirect	(Q) _____	(W) _____
Total	_____	
Total Manufacturing Cost Per Unit	_____	

Table 8-3. Summary Of Production Costs As Cutting Fluid Application Is Investigated

	Example A		Example B		Example C		Example D		Example E		Example F	
Associated, Direct												
Labor	0.264		0.261		0.259		0.259		0.251		0.193	
Cutting-Fluid	0.0143		0.0062		0.0105		0.01		0.01		0.01	
Subtotal		0.2783		0.2672		0.2695		0.269		0.261		0.203
Associated, Indirect												
Maintain Fluid System	0.011		0.0022		0.0022		0.0022		0.0022		0.0022	
Haul-away dirty fluid	0.038		0.0091		0.0026		0.0026		0.002		0.002	
Maintain pumps & piping	0.0022		0.0022		0.0022		0.0033		0.0033		0.0033	
Subtotal		0.0512		0.0135		0.007		0.0081		0.0075		0.0075
Affected, Direct												
Cutting tools	0.091		0.091		0.091		0.091		0.055		0.055	
Subtotal		0.091		0.091		0.091		0.091		0.055		0.055
Affected, Indirect												
Set-Aside for Investment	0.10		0.1016		0.1016		0.1016		0.10		0.12	
Machine tool maintenance	0.011		0.011		0.011		0.011		0.011		0.011	
Inspection tool replacement	0.002		0.002		0.002		0.002		0.002		0.002	
Allowance for scrap	0.006		0.006		0.006		0.006		0.006		0.006	
Fixed Mfg. Overhead	0.06		0.06		0.06		0.06		0.06		0.04	
Purchase raw material	0.21		0.21		0.21		0.21		0.206		0.206	
Subtotal		0.389		0.391		0.391		0.391		0.385		0.385
Total cost per piece		0.8095		0.7627		0.7585		0.759		0.7085		0.6505
Projected pieces per year		500,000		505,000		510,000		510,000		525,000		682,500

CHAPTER 9
TROUBLESHOOTING CUTTING AND GRINDING FLUID APPLICATIONS

As discussed in Chapter 4, no selection guide for machining and grinding fluids is so accurate that an exact formula will automatically provide maximum performance in a given operation. Troubleshooting is needed not only to bring an operation up to an acceptable level, but it is also helpful in achieving performance gains in what appear to be reasonably satisfactory operations.

In troubleshooting machining and grinding fluids, it is important to remember that the dual functions of lubricating and cooling are somewhat opposed. For instance, straight oil compounds will provide better antiweld and better lubricating properties than water miscible compounds, while water miscible compounds will provide better cooling. Water miscible compounds with heavy creamy emulsions containing extreme pressure additives will generally give better lubrication but poorer cooling than highly detergent chemical compounds.

Manufacturers of machining and grinding fluids are constantly trying to build optimum values of lubrication and cooling into a single fluid. While this balance of properties is highly desirable, there will always be operations where increased lubrication will eliminate problems, and others where increased heat transfer will be the answer. The direction in which to move will be established by the results obtained while evaluating fluids in the operation and the types of problems encountered.

MACHINING OPERATIONS

In most cutting operations, the cutting fluid must meet a range of operating requirements. For instance, the dwell time at the end of a cycle on an automatic screw machine creates a condition where the tool rides over the metal and gives a burnishing effect that requires some hydrodynamic lubrication. This type of condition favors straight oil lubricants, whereas the use of new rigid machines, carbide tools, and high speeds generally favor water miscible fluids. Arriving at the best solution for any problem requires an open mind and a willingness to try different fluids. This open-minded approach is needed for both troubleshooting and general improvement in overall operating performance.

Chip Welding/Tool Seizure

Excessive buildup on the tool, or seizure, can be caused by either inadequate heat dissipation or lack of antiweld qualities in the cutting fluid. This problem can be eliminated by changing from straight oils to water miscibles and, at other times, by changing from water miscibles to straight oils. The choice of fluid will depend on observing the conditions of the particular operation.

Increased heat dissipation is called for by:
1. Blued chips, burned or discolored edges of cutting tools, or hot workpieces
2. High cutting speeds
3. Machining parts with thin-wall sections
4. Continuous tool contact as in turning or drilling

Where these conditions appear in an operation, increased heat dissipation can be achieved by:
1. Proper coolant application in terms of direction and volume of flow
2. Changing from high viscosity oils to low viscosity oils to obtain better heat exchange properties
3. Changing from straight oils to water miscible fluids
4. Changing from heavy-bodied, milky emulsions with extreme pressure additives to semi-chemical or chemical types if the problem occurs when using a water miscible fluid
5. Decreasing concentration, using more water

One of the best examples of improvement in an operation through increased heat dissipation occurred in the development of high-speed tapping. Tapping had generally been carried out at slow speeds and, for years, was generally performed with straight oil compounds. Even today, on slow speed operations, straight oils will give the best performance. As tapping speeds increased, tap breakage was a frequent problem. In many instances, the problem was eliminated by converting from straight oils to water miscible fluids--the improved heat dissipation eliminated tool seizure and greatly increased tool life.

There are many cases where chip welding and tool seizure cannot be eliminated by increasing heat dissipation but require increased lubrication when any of the following conditions are present:
1. Materials are high tensile steels or heat-resistant alloys.

2. Machine speeds are low
3. Tool contact is intermittent, as in hobbing or broaching where the opportunity for heat exchange is better than in turning
4. Where it is difficult to get the lubricant to the point of cut. Conventional threading, in which the cutting edges of the chasers are supplied by only one coolant line, is a good example of an operation where inadequate lubricant application favors the use of oil-base fluids

Lubrication properties can be increased by these steps:
1. If a low or nonactive sulfurized oil is used, increase sulfur activity
2. If cutting fluid carry-in is a problem, change from a low-viscosity to a high-viscosity oil with better "cling" characteristics
3. Change from water miscible fluids to straight oils
4. Change from ordinary emulsion or chemical types of water miscibles to the heavy-bodied milky emulsions or chemical types with extreme pressure additives
5. Decrease water content; use more concentrate

Most troubleshooting operations with machining fluids center on the search for an optimum balance of lubrication and cooling. A different balance of properties is called for, not only by tool seizure or chip welding, but by changes in speeds, feeds, and tooling materials. Changes are indicated whenever speeds are too slow, tool life is short, or the finish is poor.

Poor Finish

While severe chip welding or tool seizure will always produce poor finish, there are a number of cases where inadequate microinch finish will result in an operation without evidence of excessive buildup on the tool. In many of these cases, changes in cutting fluid can yield significant gains in performance.

Where the finish on an operation is not up to standard, it is essential that an analysis be made to separate the tooling problems from the cutting fluid problems. If the finish of the first few pieces machined with sharp tools is adequate, then the problem can be approached from the cutting fluid viewpoint. If the finish of the first pieces machined with sharp tools is inadequate, then the problem must be approached from both the tool and cutting fluid viewpoint.

Where finish is inadequate at the very beginning of the operation, these points should be checked:
1. Increase chemical activity to prevent buildup on the cutting edge of the tool

2. If the operation requires carry-in of the cutting fluid, as in ID broaching, increase the viscosity and the "cling" of the cutting oil
3. If excessive heat is evidenced by discolored chips or hot workpieces, change to a lubricant with better heat exchange characteristics

In troubleshooting finish problems, it is absolutely essential that the finish problem be measured in relation to the tool condition. Problems that occur with a set of sharpened tools will call for a different course of action than problems that occur as tools wear.

Excessive Tool Wear

With the wide variety of tooling materials available today, many wear problems can be solved by a change in tool materials. While tool material is extremely important, the cutting fluid can make a significant contribution to reducing tool wear and, in some cases, can also be a contributing factor to excessive tool wear. Rapid tool wear can be caused by excessive chemical activity in the cutting fluid. Excessive chemical activity creates chemical wear by the reaction between the chemical elements in the fluid and the tooling material. Chemical wear is more frequent on operations such as broaching and tapping where tool geometry requires the use of a tool with as little back rake as possible.

In general, the attack on wear problems with cutting fluids will follow the same pragmatic approach used in attacking chip welding or tool seizure problems. At high speeds with soft metals, the emphasis should be on increasing heat dissipation qualities by changing to water miscible fluids or straight oils with better heat exchange properties. Where operations are slow and metals have high yield strength, the emphasis should be on improved lubrication.

There are no absolutes in improving tool life or finish with cutting fluids. An example is broaching where the conditions of slow speeds and high finish requirements generally favor the use of straight oils. However, on ID and spline broaching, the heat exchange problems are so severe that many creamy, extreme pressure water miscible fluids at concentrations of 5 or 7 to 1 will far out-perform straight broaching oils.

Chatter

Chatter is generally caused by inadequate rigidity in a machining operation. In some cases, adequate rigidity is difficult to achieve by mechanical means due to the design of the part being machined, or equipment or fixture problems. Where this occurs, cutting fluid can reduce or eliminate chatter by hydraulically dampening vibration and, in effect, contribute to the rigidity of the tooling setup.

The increase in tooling rigidity that can be provided by higher viscosity cutting fluids is due to the fact that the cycle time in many cutting operations does not permit the cutting fluid to flow out of the crevices and clearances in a tooling setup. In this way, the cutting fluid can make a major contribution to the rigidity of an operation.

GRINDING OPERATIONS

Grinding, like cutting, requires a combination of lubrication, cleanliness or detergency, and cooling for effective finish and long wheel life. Many engineers attempt to overemphasize the importance of cooling and cleanliness in grinding operations and thus pass up the gains that can be achieved through better lubrication.

The value of lubrication in a grinding operation can best be seen in form and thread grinding where straight oils are absolutely essential for adequate maintenance of wheel form. Too frequent wheel dressing is a sign that the grinding operation needs a softer wheel or greater lubricity or both. An inability to maintain dimensional tolerances indicates a need for greater lubricity.

In improving lubrication on a grinding operation, these steps can be taken:
1. Change from water miscible coolants to straight oils; straight oil provides maximum lubrication in grinding
2. Increase concentration from the conventional 50 to 1 down to 5 or 10 parts of water to 1 of the water miscible grinding fluid
3. Change from plain chemical or semichemical fluids to heavily compounded extreme pressure types

In following these recommendations, the greatest operational difficulty in grinding operations will occur with the creamy, highly compounded water miscible fluids. These products generally provide inadequate detergency and will create a cleanliness or chip disposal problem. The lack of detergency will also cause wheel loading and, if the area of wheel contact is large, burning. However, on cylindrical grinding operations where the wheel contact area is small, the highly compounded, creamy emulsions will give excellent performance where it is desirable to prevent wheel breakdown in order to maintain dimensional tolerances.

Burned or discolored work will generally result from pick-up of swarf or combinations of swarf and free oil on the wheel surface. This will appear as a black deposit, and the wheel, on visual inspection, will appear dirty. Where this type of pick-up and burning develops, it can be eliminated by changing to a more detergent water miscible grinding fluid or to a straight oil. Straight oils will keep the grinding wheel clean and free cutting--a requirement for good wheel life and freedom from burning.

Straight oils achieve wheel cleanliness because there is no dispersed phase as there is in an emulsion system where oil particles are distributed through a continuous water phase. With straight oils, it is possible to achieve completely clear oil base solutions of chemical additives; there is no separate phase to load the wheel. In addition, straight oils have good wetting properties and greater lubricating effect and these properties add to wheel cleanliness.

Despite the advantages of excellent wheel life, good finish, and free cutting action offered by straight oils, most operating plants prefer water miscible grinding fluids because they eliminate fire hazards and oil mist which create operator complaints (they present a health hazard) and housekeeping problems. Fire hazards with straight oils can be reduced by placing additional coolant nozzles below the grinding wheel to quench the spark train and wash away the accumulations of fine chips which provide a wicking action and create a condition that makes a fire possible. Oil mists can be handled by exhaust systems or precipitrons.

While wheel life or wheel form is an important cost consideration in a grinding operation, most plants center their attention on selecting a grinding fluid that prevents recirculation of fines or provides satisfactory filtration and meets the general finish requirements of the grinding operation.

Microinch Finish

Finish in a grinding operation is primarily a function of grinding wheel selection. However, with a given grinding wheel, it is possible to change finish by changing from one grinding fluid to another.

In many grinding operations, finish is obtained by a partial loading of the wheel. While this is a condition that in surface grinding produces burning, in straight cylindrical grinding it can help to obtain improved finish. An example of this can be seen where finish is improved by purposely contaminating a water miscible grinding emulsion or solution with a straight oil. The straight oil plates out of the solution, loads the wheel, and contributes to improved finish.

Where finish is a problem, lower microinch finishes can be achieved by:
1. Increasing the viscosity of the oil when grinding with straight oils. Higher viscosity oils will cushion a wheel and improve finish
2. Changing from a water miscible fluid to a straight grinding oil
3. Increasing the concentrations of water miscible fluids from the conventional 50 to 1 down to 5 or 10 to 1
4. Changing from a detergent type water miscible fluid to the emulsion or chemical type with extreme pressure additives

In troubleshooting grinding operations, operator acceptance is as important a consideration as satisfactory finish and wheel life. The misting of the lubricants creates odor problems, contributes to housekeeping problems and, by keeping operators' hands and arms wet, makes skin sensitivity an important factor.

SYMPTOMS, CAUSES, AND CORRECTION OF PROBLEMS

This section is concerned with the symptoms, causes, and correction of cutting fluid problems from the functional, maintenance, physiological, and economic aspects (1). See Tables:
9-1. Machining Operation Problems
9-2. Grinding Operation Problems
9-3. Maintenance Problems
9-4. Physiological Problems
9-5. Economic Problems

The success of troubleshooting any problem depends upon the gathering of all the facts and evidence contributing to the problem and arranging them logically to reach a solution. Detailed evidence must be gathered while the situation exists since little can be deduced from hearsay information later.

Machining Operation Problems

A wide variety of problems involving tool life, surface finish, and machining tolerances may occur in any machining operation. These problems are normally a function of elements in the machining process other than the cutting fluid, if the cutting fluid selection was made with a reasonable amount of judgment. The exceptions to this broad generalization usually involve materials of very poor machinability or operations with a high degree of severity where cutting fluids may often produce a functional effect on the same order of magnitude as other machining process parameters.

In general, however, for most machining operations the cutting fluid can be considered to be a *machining aid* rather than a possible source of functional machining problems. For this reason it is recommended that, when machining problems occur, all basic machining elements including cutting speed, feed, tool material and geometry, dimensions of cut, and set-up rigidity be thoroughly examined for correctness *prior* to troubleshooting the cutting fluid. No cutting fluid has yet been developed that will cure any malfunction caused by improper machining procedure.

For machining difficulties which can definitely be established as being a function of cutting fluid application, Table 9-1 can be used to determine the appropriate cause and correction of the problem.

Grinding Operation Problems

In analyzing grinding problems, the same general comments apply as in machining problems. Wheel breakdown, loading, and workpiece surface should be analyzed from the standpoint of basic grinding parameters prior to troubleshooting the fluid.

It is generally recognized, however, that fluids can produce a highly significant functional effect in grinding processes. This is particularly true when grinding very hard or heat-resistant materials. For most materials, and particularly those just mentioned, compounded oils will produce better surface finishes and less wheel wear than can be achieved with water-base fluids. In fact, most problems involved in grinding operations which employ water-base fluids can be eliminated by the simple expedient of changing to a grinding oil, providing that other process parameters are reasonably compatible with the specific grinding operation.

Often, however, this simple technical solution is complicated by economic, housekeeping, or personnel considerations and a compromise solution must be employed. One solution to a situation which requires an oil but which must employ a water-base fluid is to use a highly-concentrated mix (5:1) of a heavy duty type water miscible product.

Miscellaneous Problems

Many problems associated with cutting fluid application are of a maintenance, physiological, or economic nature rather than functional. These are mainly comprised of such problems as:
1. Corrosion, stain, or deposits on machine tools or workpieces
2. Dissolution of paint or lubricants on machine tools
3. Presence of bacteria and fungi (mold and yeast) resulting in foul odors
4. Separation of solutions or emulsions
5. Excessive foaming of fluid

It should be emphasized that any cutting fluid which is properly applied and supported with good housekeeping and maintenance programs should exhibit none of the tendencies listed above. More specifically, any cutting fluid which is properly applied under good housekeeping conditions and which produces any of the problems listed above should be replaced with a different fluid selected carefully from the thousands available. Manufacturers and suppliers usually maintain continuous testing programs and generally can supply additional assistance to the user. The maintenance, physiological, and economic problems are normally more self-identifiable.

For example, the use of water of improper hardness or other minerals with emulsifiable oils may cause an inverted mixture and definite machining problems. However, the same effect may also be caused by improper mixing, which is a maintenance problem and the effects noted will be both physiological and economic.

Examination of Tables 9-1 through 9-5 is recommended for determining side effects in various areas other than the principle problem causing the investigation.

Table 9-1. Machining Operation Problems

Symptom	Cause	Correction
Low tool life caused by abrasive wear. Chip curl decreases without burning; smoke or steam begins or increases; cratering or rapid flank or primary wear occurs.	Lack of lubrication at cutting edge of tool allows mechanical abrasion between tool and chip or tool and workpiece.	Select fluid with better lubricating qualities. Increase concentrations on water miscible fluids. Replace all old cutting oils or water miscibles. Increase volume and pressure, and direct into clearance angles.
Low tool life combined with excessive heat. Pounding in the cut; underside of chip is rough and torn; chip particles, burns, or heat checks on workpiece; chips discolored; steam, smoke, or fire; heavy BUE (built-up edge) or burns on tool.	Inadequate cooling at cutting edge of tool encourages tool seizure or build-up.	Select fluid with higher cooling capacity. Change from oil to water miscible. Increase flow or pressure; keep tool flooded. Improve application in clearance and back of chip.
Low tool life due to welding, chipping, and complete failure. Heavy BUE on tool with particles on part; chipped tool.	Lack of anti-weld properties promotes the development of a BUE type chip. The unstable BUE and the fragments leaving it are relatively much harder than the basic work material. This promotes rapid wear through siding contract, and chipping by pulling out some of the base material when it leaves.	Change to a chemically active fluid, preferably oil or one with increased EP (extreme pressure) or lubricating qualities. Replace EP oils if indicated, because of reduced chemical activity.

Table 9-1. Machining Operation Problems – Cont'd

Symptom	Cause	Correction
Low tool life due to chipping.	Intermittent or non-uniform cooling effects.	Increase flow or coverage; keep tool flooded continuously. Improve direction of application.
		Change to fluid having lower cooling properties.
		Remove fluid on carbide tool.
Tool seizure on drill, tap, reamer. Tool sticks, breaks.	Excessive thermal expansion due to inadequate lubrication of rubbing surfaces.	Select fluid, preferably oil, with better EP and lubricating qualities.
		Improve volume, pressure, and direction of fluid application.
		Check for dilution of fluid.
		Increase concentration of water miscible fluids if necessary.
Surface finish poor – rough, smeared, torn, marred, discolored.	Lack of lubrication.	Select fluid with better lubricating qualities to reduce BUE. Check for dilution of fluid.
	Chip interference.	Improve volume, pressure, and direction to move chips out of cut area.
		Improve filtration to remove particles.
Out of tolerance parts.	Non-uniform or inadequate cooling.	Keep tool and workpiece continuously flooded.
		Maintain fluid at constant temperature. Check for dilution of fluid.

Table 9-2. Grinding Operation Problems

Symptom	Cause	Correction
Loss of form; high ratio of wheel wear to metal removed.	Lack of lubrication from fluid.	Change wheel to higher density, smaller grit. Increase lubricity of coolant.
Loss of part size; frequent wheel dressing.	Lack of cooling from fluid.	Reduce viscosity. Improve volume, pressure, and direction. Reduce fluid temperature to ambient. Change to softer wheel.
Burning, warping, expansion, softening, or re-hardening of hardened steels; build-up of residential stresses; presence of steam or smoke.	Lack of lubrication.	Change to EP chemical. Decrease wheel density; increase wheel speed or decrease work speed.
	Water miscibles.	
	Oils.	Change to softer wheel and more EP fluid.
	Lack of cooling.	Increase water content, pressure, or volume.
	Water miscibles.	Apply through wheel.
	Oils.	Change to softer wheel or more cooling fluid.
Poor surface finish.	Inadequate lubrication or cooling.	Change wheel to finer grit. Increase concentration, pressure, volume. Improve filtration.
	Water miscibles.	
	Oils.	Change to fat or EP fluid. Increase volume, pressure. Improve filtration.
Wheel loading—slick, shiny, or glazed surface.	Water miscibles and oils.	
	Wheel too hard or speed too high. Grains are dull without being torn from wheel.	Use softer, more open wheel. Dilute concentration. Improve filtration.

TROUBLE SHOOTING CUTTING AND GRINDING FLUID APPLICATIONS

Table 9-3. Maintenance Problems

Symptom	Cause	Correction
Corrosion of Machines or work. Rust (2).	Exhausted or missing inhibitor. Continued makeup with water only.	Add inhibitor to water miscibles. Do not use chromates or dischromates. Check concentration.
Rust on parts.	Condensing water vapor when temperature drops below dew point, normally overnight.	Oil parts with rust preventive or preservative oil – not machine oil through which water will penetrate.
Rust, rapid.	Acidic vapors from nearby pickling tanks. Also sea air and fly ash.	Use rust preventives specially recommended to withstand these conditions.
Rust, fingerprints.	Inorganic salts in the sweat gland residues are insoluble in oil.	Must be removed by water or water-based rust preventive solutions immediately after handling. Then oil with preservative oil.
Stains, black sulfur.	Iron combines with moisture and sulfur as a black, "smelly" sulfide. Often found where parts contact each other.	Wash off oil after machining or add sulfur inhibitor to cutting fluid. If severe, remove fluid with high moisture content and replace with fresh. If water miscible, check for sulfate reducer bacteria. Replace with fresh coolant after proper cleaning of machine.
Stains, general.	Chemical reaction between cutting fluid compounds and machine or workpiece.	Check mosture content in oils. Check EP compatibility with machine or workpiece.

Table 9-3. Maintenance Problems—Cont'd

Symptom	Cause	Correction
Deposits on slides and ways; electrical shorts in motors.	Evaporation from water miscible fluids may leave hard, soft, or gummy deposits either in redissolvable or undissolvable forms.	Change to new fluid after cleaning machine. Change to de-ionized water. Eliminate fluid types with hygroscopic deposits. Specify water resistant insulation.
Paint lift or peel.	Chemical action of additives in fluid, i.e., alkali, wetting agents, etc.	Change fluid or paint.
Foaming.	Presence of wetting agents: fluid made for harder water.	Add anti-foam agent; select new fluid.
Inverted emulsions.	Improper mixing of water into oil.	Mix oil into water with continuous agitation.
Slimy oil layer on top of water.	Gradual evaporation of water until situation becomes unstable and solution inverts; or tramp oil accumulation creates water shortage.	Check concentration on scheduled periodic basis. Drain, clean, and replace with fresh mixture.
Insoluble soap curds.	Proper mixing but water is hard.	Use de-ionized water.
Frequent machine cleaning. Rapid rise of rancidity with bad odor, often over weekend.	Bacteria, algae, and fungi may develop in machines using water miscibles; incorrect or incomplete machine cleaning procedures.	Proper maintenance procedures should include thorough machine cleaning with compatible germicidal cleaner after removal of old fluid. Consider changing type of fluid.
Plugged lines. Reduced tool life.	Bacterial, fungi, and algae growth can be extremely rapid in warm fluid, with resulting chemical breakdown and creation of sludge.	All deposits, particularly in corners and pockets, should be removed mechanically, including cleaning of valves, lines, fittings, and pump. Circulate cleaner solution for sufficient time.

TROUBLE SHOOTING CUTTING AND GRINDING FLUID APPLICATIONS 175

Table 9-3. Maintenance Problems—Cont'd

Symptom	Cause	Correction
		Drain cleaner and sediment from sump. Circulate with clean water for sufficient time and drain.
		Refill machine with clean fluid properly mixed in clean containers using clean, de-ionized water.
		Maintain good housekeeping during use. Check mixture concentration regularly and makeup as required.
Fire or heavy smoke.	Low flash point cutting fluids.	Change cutting fluids to heavy-duty, EP type.
Fuming or boiling in reservoir or temperature.	Chemical reaction of chips or fines with cutting fluid. Common with nitrite solutions and aluminum.	Select new fluid. Improve filtration.
Weakening of grinding wheel bond.	Chemical action of cutting fluid.	Select new wheel or grinding fluid.
Chips and fines in fluid.	Inadequate filtration systems.	Check filters and strainers for proper functioning and establish periodic check.
		Severe cases may require additional facilities for centrifuging, or the use of settling tanks.

Table 9-4. Physiological Problems

Symptom	Cause	Correction
Dermatitis—operator's skin is dry and cracked, has rashes, pimples, or burns or is irritated by general inflammation or sores.	Defatting of skin by solvents or low viscosity petroleum products.	With low viscosity fluids, prevent skin contact by providing coverage of exposed areas (gloves, gauntlets, etc.). Be especially careful when operator or set-up man goes from cutting oils to water miscible coolants (or vice versa); intermixing of fluids on skin is usually irritating.
	Excessively high alkalinity of some water miscible fluids.	When using straight oils, use water soluble barrier cream on hands and arms if necessary. When using water miscible fluids, use a water impervious barrier cream. Germicides in some creams can be irritants.
	Breakdown into acids of some chlorine containing cutting oils either by moisture of the skin or when such fluids get into water miscible mixtures, e.g., brush-on oils used on turret lathes where water miscible fluid is in the sump. Or, oil on parts from previous operation carried over into a water-mix fluid.	All fluids become contaminated with use. Establish periodic cleaning of machines, sumps, and lines, with germicidal cleaner, followed by satisfactory rinsing and replenish with fresh fluid. Do not permit refuse such as food, tobacco, or sputum to get into coolant.
		Personal cleanliness is the most important preventive to dermatitis. Instruct operators in proper cleaning of hands, arms, face, and use of clean clothing.
		In the rare cases of operator allergic reaction, assignment to other tasks may be necessary.
NOTE: Dermatitis is usually *not* caused by bacteria but is usually chemical in origin.		
Odor.	Bacteria contaminated water miscible cutting fluids.	Establish and maintain regular schedule of preventive maintenance. Drain old fluid, clean machine, and replace with fresh fluid.
	Chemical compounds in fluids or chemical reactions between fluids and workpiece.	Exhaust air insufficient in volume. Change to new fluid.
Mist or fog.	Air/liquid mist systems.	Increase air exhaust systems. Install automatic cutoff to prevent continuous spray during non-cutting periods. Change to flood application.

TROUBLE SHOOTING CUTTING AND GRINDING FLUID APPLICATIONS

Table 9-5. Economic Problems

Symptom	Cause	Correction
Maintenance expenses; basic cost of fluids; price per gal.	Initial machining loading.	Cutting fluids are an expendable item and should account for a small per cent of the total manufacturing cost.
		The lowest price per gal. will not always produce, under analysis, the lowest cost per cu. in. of metal removed.
	Replacement of fluids.	The continual loss of fluid coating the chips may, under heavy usage conditions, be minimized by such mechanisms as storage over drip pans, heating, crushing, or centrifuging.
		Evaporation of water from water miscibles requires periodic replacement.
		Loss of active elements and chemicals occurs more slowly and is augmented by periodic replacements, although the life of water miscible is generally 6 months and the oils should be replaced, depending upon type and usage, usually within a 6 to 18 month period.
Evaluation of what price to pay.	Large selection of fluid types and suppliers.	"The cost of the fluid per part will normally be completely negligible compared to the machining cost per part and that fluid which gives the smallest machining cost per part is normally the most economical regardless of the cost of the fluid concentrate. Thus it is clear that if a particular fluid really does a superior job and the machining cost per part is reduced any measurable amount by using it, this water base fluid should be used regardless of whether the concentrate costs one, ten or twenty dollars per gallon" (3).

178 SYMPTOMS, CAUSES, AND CORRECTION OF PROBLEMS

Table 9-5. Economic Problems–Cont'd

Symptom	Cause	Correction
Evaluation of what price to pay.	Large selection of fluid types and suppliers.	"In most cases, cutting fluids actually *set the limit* on what cutting tools can do in respect to finish, tool life and production rate. Consequently, cutting fluids are definitely a significant production item and should be chosen not on a price per gallon but on the ability to reduce unit production cost" (4).
High control or maintenance labor costs.	Unit machine maintenance labor, or production interference.	Evaluate the possible utilization of a central system where product and fluid mix is relatively uniform. Maintaining the proper pH balance of water miscible and the chemical activity, viscosity, temperature, reconditioning, and filtration of all fluids may prove more economical with total and partial central systems.
Storage and handling costs.	Fluids are shipped generally in sterile, heavy-duty 55 gal. drums that require minimum protection before being opened. Normal costs arise from warehousing or storage sufficient to protect from excessive heat or cold (minimum of 50°F.) which may cause chemical or physical changes in either the active oils or water miscible concentrates.	Provide general warehousing before opening, or if stored outside, lay the drums on their sides. After opening, maintain separate clean area under control of trained personnel to prevent improper mixing of water miscibles or intermix of types of cutting fluids.

REFERENCES

1. ADP Machinability Laboratory, Lockheed-California Company, Burbank, California, (1966).
2. E. W. Hitchcock, "Are You Producing Rust-Free Parts?", Rust-Lick, Inc., (1965).
3. M. C. Shaw and P. A. Smith, "Methods of Applying Cutting Fluids," *ASTE Research Report No. 3*, (1956).
4. C. A. Sluhan, "The Effective Utilization of Cutting Fluids to Improve Metal Removal Rates," *ASTME Technical Paper, SP 63-188*, (1963).

CHAPTER 10
CREEP-FEED GRINDING AND GRINDING WITH SUPERABRASIVES

Creep-feed grinding and grinding with superabrasives represent an expanding metalworking technology and have special considerations. The material below goes beyond the basic material on grinding provided in Chapters 3.

CREEP-FEED GRINDING

Creep-feed grinding is almost always performed in the climb grind mode, as shown in Figure 10-1. The maximum amount of heat is generated at the top of the arc of cut and decreases to a minimum at the bottom. In the climb grind mode, the coolest cutting fluid and the sharpest grinding wheel enter the arc at the point of maximum stock removal and highest energy. The cutting fluid is generally applied to the process at the top of the arc allowing the porosity within the grinding wheel to transport the fluid into and around the arc of cut. It is imperative that the cutting fluid is applied properly to provide the necessary cooling and lubrication in the arc of cut. Unfortunately, creed-feed grinding processes in the industry suffer from poor application of the cutting fluid. Creep-feed grinding has an inherently long arc length of cut and it has been proven that the cutting fluid, even when applied properly, warms up as it moves around the arc of cut. That means that there is a physical limitation on the heat transfer capacity of the cutting fluid.

In order to maximize the benefits of the cutting fluid, it is important to design a cutting fluid system which is able to deliver cutting fluid at a rate of 1.5 to 2 US gpm for every unit of horsepower used in the grind. For example, for a creep-feed grinding machine with a maximum of 50 hp (37.3 kw) the cutting fluid system should be capable of delivering 75 to 100 US gpm. To ensure that the fluid is applied properly, the speed of the fluid needs to be equal to, or slightly greater than, the peripheral speed of the grinding wheel. There are 231 cubic inches ($3.8m^3$) in a US gallon, so a 100 US gpm system will supply 23100 in^3 of fluid per minute. If the grinding

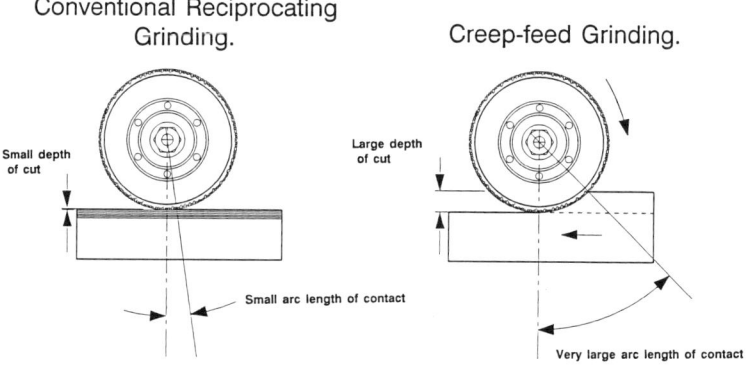

Figure 10-1. *Creep-feed grinding.*

wheel is to be run at 6000 sfm (30ms^{-1}) and the wheel is 2 inches wide (50.8mm), the volume flow of fluid will need to be (6000 x 12) inches per minute, multiplied by the area of the nozzle. To be sure that the nozzle covers the full width of the wheel, the nozzle might be say, 2.25 inches (57.15 mm) wide. The height of the nozzle opening (d inches) can be found from the follow equation:

The pump delivers 23100 in^3/min.
The volume flow (6000 x 12) x 2.25 x d in^3/min.
Therefore 23100 = (6000 x 12) x 2.25 x d
d = 0.1426 inches

If the nozzle is made 2.25 by 0.125 inches (50.8 by 3.1 mm) then the speed of the fluid will be between 10 and 15% faster than the grinding wheel periphery, which is ideal. The extra speed makes up for the drop in velocity between the exit from the nozzle and the point at which the fluid impinges the wheel.

Creep-feed grinding also requires a cutting fluid dam to support the fluid flow through to the end of the cut, as indicated by Figure 10-2. Without the dam, there is a high risk of coolant starvation as the wheel exits the workpiece. When starvation occurs at the end of a cut, heat builds up in the surface, the material expands into the wheel and results in a ramp on the end of the part.

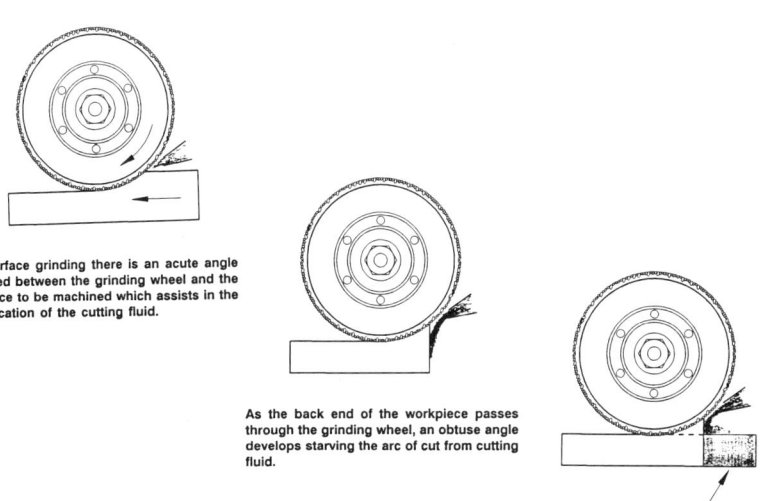

Figure 10-2. *Fluid dam.*

Creep-feed grinding generates very low forces on each individual grain of the grinding wheel periphery. The wear factor is therefore significantly lower than that of conventional grinding processes. Creep-feed grinding wheels are porous and quite soft, allowing a very sharp wheel periphery to be presented to the surface of the part. In continuous dress application of creep-feed grinding, the grinding wheel is held in a controlled state of sharpness throughout the process. It is generally accepted that due to the inherent sharpness of a creep-feed grinding wheel, the process suffers more from a breakdown in cooling than a lack of lubrication. Lubrication becomes more important as materials become more difficult to machine because of their hardness, "gumminess" or crack sensitivity.

Refrigeration of the cutting fluid is necessary in creep-feed grinding, not to keep the process dimensionally stable, but to control the heat transfer of the cutting fluid. The grinding action generates heat, but there is also a great input of energy from the pumping and the churning of the fluid, which dramatically increases the temperature of the fluid. Creep-feed grinding machines, without refrigeration, typically run the cutting fluid up to and over 100°F (37.8°C) which significantly detracts from the effective cooling capability of the fluid.

The creep-feed grinding process can generate large amounts of heat sufficient to change the heat transfer characteristics of the fluid. This is seen in the switch from nucleate to film boiling, as shown by Figure 10-3. This switch occurs for both water-based fluids as well as straight oils, but at different temperatures. Nucleate boiling is the very efficient heat transfer type of boiling which takes place at the workpiece surface, with good fluid agitation, good convection and good conduction of the heat energy into the fluid. Once the heat intensity at the workpiece surface reaches a level high

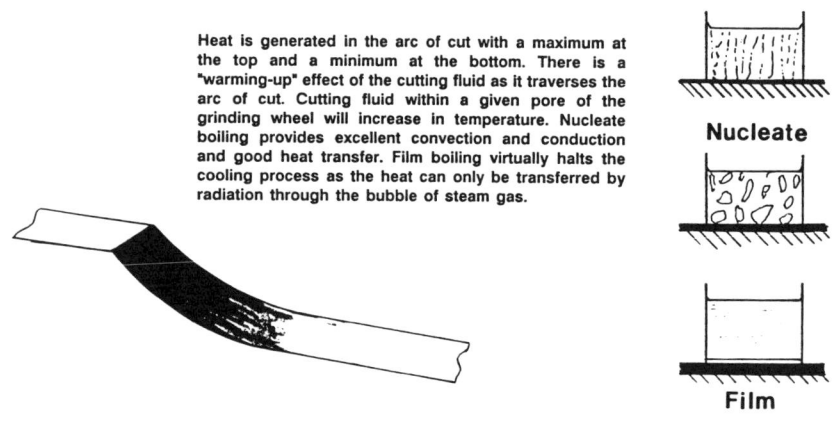

Figure 10-3. *Nucleate to film boiling.*

enough to cause film boiling to occur, the fluid separates from the surface. The cutting fluid is then supported on a film or bubble of steam gas in the case of water, and oil vapor in the case of an oil, preventing both conduction and convection from taking place. The cutting fluid can no longer work as a coolant so the heat is channeled directly into the workpiece surface. When this occurs, the material is generally metallurgically damaged and usually shows visible signs of burning.

GRINDING WITH SUPERABRASIVES

Grinding with superabrasives falls into two categories: Diamond and CBN. Grinding with diamond grinding wheels is becoming more popular with the increase in demand to grind ceramics. Ceramics are generally very hard and brittle materials which are prone to chipping, cracking and poor surface integrity.

Grinding with diamond requires high lubricity more than cooling capability. Diamond is used up to 9000 sfm (45ms^{-1}). Higher wheel speeds generate too much frictional energy and cause high wear on the diamond abrasive. Resin bonded diamond wheels are the most popular in the industry with metal and plated bonds a close second and third. These bond systems exhibit little to no porosity and therefore have no room for large volumes of fluid or chip clearance in the arc of cut.

In order to machine ultrahard materials with diamond, the diamond has to penetrate the surface of the material or cause sufficient stress at the surface and in the subsurface to cause material to be sheared or expelled from the surface. Machine tool technology, however detracts most from the ability to grind ceramics with a high surface integrity due to their lack of stiffness. High concentration, between 10 and 20%, of high pressure additives in waterbased fluids are most suitable. Due to the tenacity of straight oils to coat the grinding wheel periphery and their high lubricity, they can significantly improve surface integrity as well as the life of the diamond abrasive. Before using an oil on a ceramic, it is necessary to be sure that the oil will not contaminate the material.

Grinding with CBN is an altogether different story. CBN is best used at very fast wheel speeds. CBN is an efficient abrasive at 12,000 to 15,000 smf (60 to 75ms^{-1}) but will become more cost effective and provide a better workpiece surface in the range of 25,000 to 45,000 smf (125 to 250ms^{-1}). This poses a significant problem in the application of the cutting fluid. Attempting to match the cutting fluid velocity with the wheel speed is very difficult. A different method of applying the fluid has to be employed. A wrap-around manifold, as show in Figure 10-4, is the most effective. This system allows the surface of the grinding wheel to pick up the fluid and accelerate it around the manifold until the fluid reaches wheel speed. The fluid is then ejected into the nip between the wheel and the workpiece surface to be machined.

Although straight oil appeared vastly superior to water-based fluids in the early days of CBN, the new chemistries of the high pressure additives and more stable emulsions and tighter bonding of the fluids, means that water-based fluid, although they will not be as good, overall, as straight oil, are fast becoming almost as good as the straight oils in grinding. They also decrease the risk, particularly at high speeds, of fire hazards, unhealthy oil mists, sticky messes and part contamination problems that can be encountered with oils.

It is also well to remember that some straight oils will degrade and soften a resin bond matrix. With the upsurge in vitrified bond superabrasive wheels this is less of a problem. The increased porosity and chip clearance of the vitrified bonds assist in good cutting fluid application for both straight oils, as well as water-based fluids, and have been major factors in the increased use of superabrasive grinding wheels.

Cutting fluid enters the manifold and accelerates around the wheel periphery between the closely fitting plate and the wheel surface. There are patented designs of ribbed plates which are said to pulse the fluid around the wheel periphery and de-load the wheel.

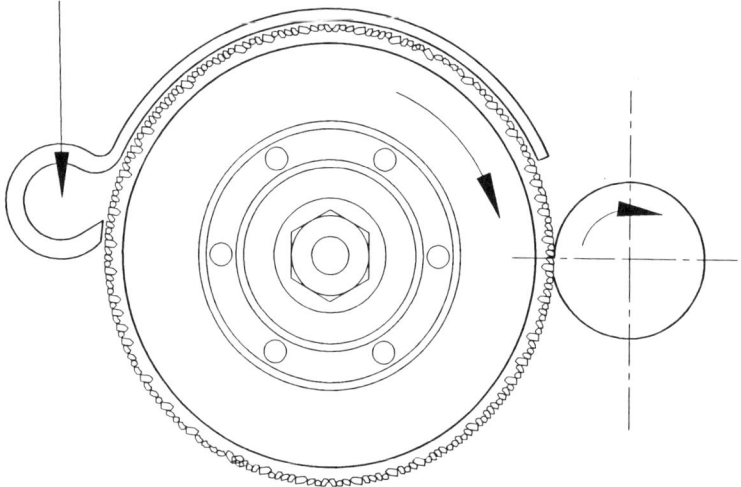

Figure 10-4. *Wraparound manifold.*

CHAPTER 11
HEALTH AND SAFETY, ENVIRONMENTAL AND REGULATORY CONSIDERATIONS OF METALWORKING FLUIDS

The areas of health and safety, environmental issues, and regulatory matters are all intertwined with one another. Each area will be discussed separately, however, overlaps will occur within the areas. Some of the regulations to be discussed are the Occupational Safety and Health Act and the Toxic Substances Control Act which were passed to protect human health. The Occupational Safety and Health Administration (OSHA) created a series of standard regulations covering personal protection, fire standards, chemical substances and both acute and chronic effects. Another regulation that will be discussed is the Resource Conservation and Recovery Act (RCRA) which is an environmental statute that was signed into law in October, 1976. RCRA falls under the jurisdiction of the Federal Environmental Protection Act (EPA) which issued final rules in the May 19, 1980, Federal Register. The effective date for these regulations was November 19, 1980. Since that time, EPA has issued over 1000 pages of additional regulations in the Federal Register in its attempt to completely enact the RCRA legislation.

HEALTH AND SAFETY

To help in the protection of human health, OSHA has set permissible exposure limits for chemicals, such as $5mg/m^3$ for oil mist, a common occurrence in the use of cutting fluids. One of the most important standards passed by OSHA is the Hazard Communication Rule (CFR1910.1200) which went into effect on November 25, 1985. This regulation educates the employee about the chemicals they work with, enabling them to use the proper equipment and understand the potential hazards of these chemicals in order to reduce occupational exposure risks.

Some of the considerations to be covered under health and safety are product selection, storage, dispensing and maintenance. Besides the exposure to a metalworking fluid as manufactured, there should be concern with exposure to both concentrated and diluted metalworking fluid additives and biocides.

Routes of exposure to metalworking fluids include dermal contact, eye contact, inhalation and ingestion. This exposure may be from direct contact or from airborne mists and residues. This contact may cause dry skin, rashes or irritation of the skin, irritation of the eyes, respiratory irritation or gastrointestinal disturbances. The manufacturer's material safety data sheet and toxicity studies on the metalworking fluid can be reviewed to determine any potential exposure problems and the necessary safety precautions.

The use of the metalworking fluid itself should be appropriate for the intended operation. The product recommendation and correct dilution should come from the manufacturer. All additives and biocides should only be used as recommended by the manufacturer. Special safety precautions should be taken with additives and biocides, as they present a greater risk to the operator than the metalworking fluid.

The metalworking fluid needs to be maintained to ensure chemical stability and operator safety. Some of the tests to maintain control are: fluid concentration, pH, bacteria and fungus levels, tramp oil, corrosion inhibitors, biocide concentrations and elemental analysis for specific components. The manufacturer should supply the customer with the specific parameters for the fluid. The results of these tests will be evaluated to determine if additional fluid or additives are required to maintain the fluid. How often these tests are performed will also be determined by the manufacturer.

Besides testing the coolant to ensure it is operating correctly, it must also be tested for any contaminants. A metalworking fluid can be contaminated with way oils, rust preventatives, cleaners and solubilized metals which can degrade the fluid. The tests mentioned above will help in determining if these have contaminated the fluid. However, they cannot determine if a metal has been solubilized so a test for metal contaminants needs to be performed on a regular basis. Suspended or dissolved metals may contribute to skin irritation and fluid degradation. These metals come from the material that is being processed. They may include nickel, cadmium, chromium and lead. It has been suggested that dissolved metal concentrations above 100 ppm increase skin irritation.

ENVIRONMENTAL CONSIDERATIONS

The manufacturer's material safety data sheet will contain specific clean up procedures for spills and leaks. Most manufacturers will also provide procedures for disposal. However, these procedures will usually be general as each city and state may have their own regulations on the disposal of metalworking fluids. It should also be noted that the material being worked and any additives used must also be considered as they affect the fluids.

The Toxic Substances Control Act, as mentioned earlier, states that a product cannot be used if its ingredients are not listed with the government. This is a way to control the type of substances which can be used in manufacturing.

Contamination of the air, water and land are some of the areas in which metalworking fluid users need to be concerned. There are many regulations to cover all these areas from the local, state and federal levels. The best way to prevent contamination is by having a good coolant management program.

RCRA gives the EPA the authority to regulate industrial waste from generation through transportation to treatment and ultimate disposal; the so called "cradle to grave" approach. Currently the major thrust of RCRA is directed toward the management of hazardous waste. EPA has indicated in its regulatory agenda that it does intend to designate waste oil (i.e., engine, hydraulic, lubricating and cutting oils) as hazardous waste. It is anticipated that this designation will include all cutting and grinding fluids regardless of composition. While no date has been set by EPA for this proposal, several states have already promulgated their own state laws which define oils as hazardous waste.

The effect this will have on a business will be increased cost in terms of disposal, administrative, labor and insurance expenses. It will also increase the amount of paperwork in terms of manifest system, labeling, reporting and record keeping. And it will increase the liability. The solution involves setting up a program that extends the life of the coolant, therefore decreasing the amount of environmental pollution and eliminating some of the cost, paperwork and liability.

Rather than disposing of the coolant, the coolant is recycled. The used coolant is removed from the machine, and the tramp oil and metal fines are removed from the dirty coolant by the recycling process. Fresh, clean coolant is then added to the recycled coolant and the new mixture goes back into the machine. The recycling process eliminates the frequency that the coolant, which is 90% water, is hauled away. This can save a substantial amount as the costs for haul away have increased considerably over the years as shown in Table 11-1. This cost will continue to increase if the EPA passes its regulation stating that used oil, which includes metalworking fluids, is hazardous. It is therefore very advantageous for the metalworking fluid user to consider some form of extending the life of the coolant.

Table 11-1. Haulaway Cost For Spent Coolant By Region

REGION	1983	1985	1988	1991
Northeast	.20 - $2.00	.20 - $3.00	.24 - $3.00	.50 - $5.00
Southeast	.20 - $1.00	.20 - $1.25	.30 - $2.00	.20 - $2.75
Midwest	.12 - $1.00	.18 - $1.00	.17 - $1.75	.22 - $3.00
Northwest	.30 - $1.00	.50 - $1.75	.70 - $2.25	2.00 - $4.00
Southwest	.07 - $1.50	.16 - $1.75	.60 - $2.00	.75 - $5.00

Note: These costs do not include transportation, lab fees, surcharges for waste

It is important when selecting a company that hauls the used coolant and treats it, to insure that they are a licensed and reputable company. If they dispose of the coolant illegally the law will fine the waste generator as the responsible party.

Besides having the used coolant hauled away, at a cost which includes paying for the disposal of water as well as concentrate, another option would be to treat the coolant for disposal. This procedure involves breaking the coolant down; the reduced volume of oil and organic material is still hauled away, but at a lower amount. The operator then treats the water phase so it can be disposed of in the sewer, and further treated by the water company. The water phase should not go into a waterway. The oil which is disposed can be rerefined.

When one discusses air pollution in connection with metalworking fluids, the matter of volatile organic compounds comes to mind. The material safety data sheet should list the fluid components with their percentages so volatile organic compounds can be determined. However, another area of air pollution is also very important. This concerns air mist and the threshold limit values to protect the operators. The values should be recorded on the material safety data sheet. These mists can be controlled by smog hogs, exhausts fans, reducing coolant nozzle spray and possible changes in feed and speed rates.

REGULATORY CONSIDERATIONS

The 1970s saw the passage of numerous legislative efforts to improve worker health and safety and protect the environment. The Occupational Safety and Health Act, National Environmental Protection Act, Clean

Water Act, Cleaning Drinking Water Act, Clean Air Act, Toxic Substances Control Act, Comprehensive Environmental Response, Compensation and Liability Act ("Superfund"), and Resource Conservation and Recovery Act are just a few. Promulgation of regulations by the Occupational Safety and Health Administration (OSHA) and the Environmental Protection Agency (EPA) to implement these laws made the chemical industry one of the most regulated industries in America.

This regulatory trend has carried on into the 1990s, but with a slightly different emphasis. Chemical user industries, including the metalworking industry, are now being more directly impacted by the regulation of chemicals. Regulations such as OSHA's Hazard Communication Standard (worker right-to-know), and the EPA's implementation of rules for the Emergency Response and Community Right-To-Know Act, place considerable responsibilities on all manufacturing sectors using chemicals in their processes, not just on the chemical makers. These responsibilities include education and training of workers, use of proper labels and material safety data sheets, written plans, maintenance of inventory records, and submission of inventory and release data.

All metalworking fluids will be covered by many of the regulations to some degree or another. This means users of cutting and grinding fluids find themselves subjected to many of these regulations. To determine which regulations apply and to what degree, takes up a considerable amount of time, but the consequences of not complying can result in fines and jail sentences.

APPENDIX

Appendix A. Machinability Ratings[1,2] Based on Cutting Speeds for 30 min. Tool Life for HSS and Carbide Tools

Material	Condition and Microstructure	BHN	HSS Tool for 30 min. life ft./min.	Carbide Tool for 30 min. life ft./min.
B 1112	Annealed 10% Pearlite - 90% Ferrite - Sulfides	135	240	920
AISI 1020	Annealed 10% Pearlite - 90% Ferrite	115	180	800
AISI 8620	Annealed 30% Pearlite - 70% Ferrite	135	182	660
AISI 3140	Annealed 75% Pearlite - 25% Ferrite	190	90	367
AISI 3140	Quenched and Tempered Tempered Martensite	302	71	340
AISI 3140 Resulfurized	Annealed 75% Pearlite - 25% Ferrite - Sulfides	190	130	485
AISI 3140 Resulfurized	Quenched and Tempered Tempered Martensite and Sulfides	296	106	305
AISI 4140	Spheroidized Spheroidized Carbides and Ferrite	166	152	700
AISI 4140	Annealed 90% Pearlite - 10% Ferrite	192	95	385
AISI 4140	Quenched and Tempered Tempered Martensite	300	69	303
AISI 4140 Resulfurized	Annealed 90% Pearlite - 10% Ferrite	190	123	400
AISI 4140 Resulfurized	Quenched and Tempered Tempered Martensite - Sulfides	300	90	310
AISI 4340	Spheroidized Spheroidized Carbides and Ferrite	206	98	480
AISI 4340	Annealed 100% Pearlite	221	88	380
AISI 4340	Quenched and Tempered Tempered Martensite	300	63	300
AISI 5140	Annealed 80% Pearlite - 20% Ferrite	192	120	362
AISI 8640	Annealed 75% Pearlite - 25% Ferrite	190	107	450

Appendix A. Machinability Ratings[1,2] Based on Cutting Speeds for 30 min. Tool Life for HSS and Carbide Tools — Cont'd

Material	Condition and Microstructure	BHN	HSS Tool for 30 min. life ft./min.	Carbide Tool for 30 min. life ft./min.
AISI 8640	Quenched and Tempered Tempered Martensite	300	72	225
AISI 8640 Resulfurized	Annealed 65% Pearlite - 35% Ferrite - Sulfides	185	135	460
AISI 8640 Resulfurized	Quenched and Tempered Tempered Martensite - Sulfides	300	102	307
AISI 52100	Spheroidized Spheroidized Carbides and Ferrite	190	123	340
14 CMV	Quenched and Tempered Tempered Martensite	341	54	375
17-22 AS	Quenched and Tempered Tempered Martensite	341	41	215
D6AC	Quenched and Tempered Tempered Martensite	56R$_c$		64
18% Nickel 250 Grade Maraging Steel	Annealed Martensite	341	92	415
18% Nickel 250 Grade Maraging Steel	Annealed and Maraged Martensite	53R$_c$	53	180
18% Nickel - 300 Grade Maraging Steel	Annealed Martensite	302	87	450
18% Nickel - 300 Grade Maraging Steel	Annealed and Maraged Martensite	54R$_c$	40	175
HP 9-4-25	Annealed Spheroidized	375	76	305
HP 9-4-25	Quenched and Tempered Tempered Martensite	415	74	300
Vascojet 1000	Quenched and Tempered Tempered Martensite	346	60	200
Vascojet 1000	Quenched and Tempered Tempered Martensite	52R$_c$	23	133
302 SS	Annealed Austenite	170	93	310
347 SS	Annealed Austenite - Carbides	175	70	280
430 SS	Annealed Ferrite - Carbides	181	160	800
410 SS	Annealed	202	158	640
410 SS	Quenched and Tempered Tempered Martensite	352	73	260
422 SS	Quenched and Tempered Tempered Martensite	352	62	300
17-7 PH	Annealed Austenite	170	70	430
Ti 8Al-1Mo-1V	Solution Treated Alpha	311	65	250
Ti 8Al-1Mo-1V	Solution Treated and Aged Alpha	341	51	210
Ti 2Fe-2Cr-2Mo	Annealed Alpha-Beta	302	43	190

Appendix A. Machinability Ratings[1,2] Based on Cutting Speeds for 30 min. Tool Life for HSS and Carbide Tools — Cont'd

Material	Condition and Microstructure	BHN	HSS Tool for 30 min. life ft./min.	Carbide Tool for 30 min. life ft./min.
Ti 6Al-4V	Annealed Alpha-Beta	312	72	180
Ti 6Al-4V	Solution Treated and Aged Alpha-Beta	365	59	158
Ti 6Al-6V-2S$_n$	Annealed Alpha-Beta	331	62	200
Ti 6Al-6V-2S$_n$	Solution Treated and Aged Alpha-Beta	429	45	175
Ti 4Al-4Mn-1V	Annealed Alpha-Beta	341	25	140
Ti 3Al-5Cr	As Forged Alpha-Beta	363	15	118
Ti 7Al-4Mo	Annealed Alpha-Beta	341	35	270
Ti 7Al-4Mo	Solution Treated and Aged Alpha-Beta	388	22	220
Ti 3Al-13V-11Cr	Solution Treated Beta	285	15	120
Ti 3Al-13V-11Cr	Solution Treated and Aged Beta	400	10	90
Inconel 700	Solution Treated Austenitic	302	18	78
Inconel 718	Solution Treated Austenitic	279	39	96
Inconel 718	Solution Treated and Aged Austenitic	45R$_c$	32	114
Inconel 901	Solution Treated Austenitic	262	28	77
Udimet 500	Solution Treated Austenitic	340	17	50
Udimet 500	Solution Treated and Aged Austenitic	360	18	56
Waspaloy	Solution Treated Austenitic	341	26	130
Waspaloy	Solution Treated and Aged Austenitic	388	23	130
Rene 41	Solution Treated Austenitic	321	17	73
Rene 41	Solution Treated and Aged Austenitic	365	13	66
HS-25	Solution Treated Austenitic	229	26	96
A-286	Solution Treated Austenitic	197	52	150
N-155	Solution Treated and Aged Austenitic	200	31	160
19-9DL	Stress Relieved Austenitic	210	39	155
Moly-TZM	Extruded and H.R. Austenitic	217		350
Moly-0.5Ti	Stress Relieved Austenitic	220		250

Appendix A. Machinability Ratings[1,2] Based on Cutting Speeds for 30 min. Tool Life for HSS and Carbide Tools — Cont'd

Material	Condition and Microstructure	BHN	HSS Tool for 30 min. life ft./min.	Carbide Tool for 30 min. life ft./min.
D-31	Extruded and Stress Relieved Austenitic	207	70	305
90Ta-10W	As Forged Austenitic	207	68	80

[1] Metcut Research Associates, Inc., Cincinnati, Ohio, 1966.
  (a) *U.S. Air Force Machinability Report*, 1950, Vol. 1.
  (b) *U.S. Air Force Machinability Report*, 1951, Vol. 2.
  (c) *U.S. Air Force Machinability Report*, 1954, Vol. 3.
  (d) *U.S. Air Force Machinability Report*, 1960, Vol. 4.
  (e) U.S. Air Force, *Final Report on Machining of Refractory Materials,* Report No. ASD-TDR-581, Contract AF 33 (600)-42349, (July 1963).
  (f) U.S. Air Force, *Final Report on Machinability of Materials,* AFML-TR-65-444, Contract AF 33 (615)-1385, (January 1966).
[2] ADP Machinability Laboratory, Lockheed-California Company, Burbank, California, 1966.

Appendix B. Approximate Relation Among Various Hardness Scales[1]

Diam mm 3,000 kg	Brinell[2] hardness Tungsten carbide 10-mm ball	A scale 60 kg Brale	B scale 100 kg 1/16-in. ball	Rockwell hardness C scale 150 kg Brale	Superficial 30 N	Diamond Pyramid hardness number, Vickers	Shore scleroscope hardness number	Approx tensile strength 1,000 psi
		86.5	.....	70.0	86.0	1076		
		86.0	.....	69.0	85.0	1004		
		85.6	.....	68.0	84.4	940	97	
		85.0	.....	67.0	83.6	900	95	
	757	84.4	.....	65.9	82.7	860	92	
2.25	745	84.1	.....	65.3	82.2	840	91	
	722	83.4	.....	64.0	81.1	800	88	
	710	83.0	.....	63.3	80.4	780	87	
2.35	682	82.2	.....	61.7	79.0	737	84	
2.40	653	81.2	.....	60.0	77.5	697	81	
2.45	627	80.5	.....	58.7	76.3	667	79	323
2.50	601	79.8	.....	57.3	75.1	640	77	309
2.55	578	79.1	.....	56.0	73.9	615	75	297
2.60	555	78.4	.....	54.7	72.7	591	73	285
2.65	534	77.8	.....	53.5	71.6	569	71	274
2.70	514	76.9	.....	52.1	70.3	547	70	263
2.75	495	76.3	.....	51.0	69.4	528	68	253
2.80	477	75.6	.....	49.6	68.2	508	66	243
2.85	461	74.9	.....	48.5	67.2	491	65	235
2.90	444	74.2	.....	47.1	65.8	472	63	225
2.95	429	73.4	.....	45.7	64.6	455	61	217
3.00	415	72.8	.....	44.5	63.5	440	59	210
3.05	401	72.0	.....	43.1	62.3	425	58	202
3.10	388	71.4	.....	41.8	61.1	410	56	195
3.15	375	70.6	.....	40.4	59.9	396	54	188
3.20	363	70.0	.....	39.1	58.7	383	52	182
3.25	352	69.3	(110.0)	37.9	57.6	372	51	176
3.30	341	68.7	(109.0)	36.6	56.4	360	50	170
3.35	331	68.1	(108.5)	35.5	55.4	350	48	166
3.40	321	67.5	(108.0)	34.3	54.3	339	47	160
3.45	311	66.9	(107.5)	33.1	53.3	328	46	155
3.50	302	66.3	(107.0)	32.1	52.2	319	45	150
3.55	293	65.7	(106.0)	30.9	51.2	309	43	145
3.60	285	65.3	(105.5)	29.9	50.3	301	...	141
3.65	277	64.6	(104.5)	28.8	49.3	292	41	137
3.70	269	64.1	(104.0)	27.6	48.3	284	40	133
3.75	262	63.6	(103.0)	26.6	47.3	276	39	129
3.80	255	63.0	(102.0)	25.4	46.2	269	38	126
3.85	248	62.5	(101.0)	24.2	45.1	261	37	122
3.90	241	61.8	100.0	22.8	43.9	253	36	118
3.95	235	61.4	99.0	21.7	42.9	247	35	115
4.00	229	60.8	98.2	20.5	41.9	241	34	111
4.05	223	59.7	97.3	(18.8)	.....	234		

Appendix B. Approximate Relation Among Various Hardness Scales[1] — Cont'd

	Brinell[2] hardness		Rockwell hardness			Diamond Pyramid hardness number, Vickers	Shore sclero-scope hardness number	Approx tensile strength 1,000 psi
Diam mm 3,000 kg	Tungsten carbide 10-mm ball	A scale 60 kg Brale	B scale 100 kg 1/16-in. ball	C scale 150 kg Brale	Super-ficial 30 N			
4.10	217	59.2	96.4	(17.5)	....	228	33	105
4.15	212	58.5	95.5	(16.0)	....	222	...	102
4.20	207	57.8	94.6	(15.2)	....	218	32	100
4.25	201	57.4	93.8	(13.8)	....	212	31	98
4.30	197	56.9	92.8	(12.7)	....	207	30	95
4.35	192	56.5	91.9	(11.5)	....	202	29	93
4.40	187	55.9	90.7	(10.0)	....	196	...	90
4.45	183	55.5	90.0	(9.0)	....	192	28	89
4.50	179	55.0	89.0	(8.0)	....	188	27	87
4.55	174	53.9	87.8	(6.4)	....	182	...	85
4.60	170	53.4	86.8	(5.4)	....	178	26	83
4.65	167	53.0	86.0	(4.4)	....	175	...	81
4.70	163	52.5	85.0	(3.3)	....	171	25	79
4.80	156	51.0	82.9	(0.9)	....	163	...	76
4.90	149	49.9	80.0	.....	....	156	23	73
5.00	143	48.9	78.7	.....	....	150	22	71
5.10	137	47.4	76.4	.....	....	143	21	67
5.20	131	46.0	74.0	.....	....	137	...	65
5.30	126	45.0	72.0	.....	....	132	20	63
5.40	121	43.9	69.8	.....	....	127	19	60
5.50	116	42.8	67.6	.....	....	122	18	58
5.60	111	41.9	65.7	.....	....	117	15	56

[1] Values in parentheses are beyond normal range and are given for information only.
[2] The Brinell values in this table are based on the use of a 10-mm tungsten carbide ball; at hardness levels of 429 Brinell and below, the values obtained with the tungsten carbide ball, the Hultgren ball, and the standard ball are the same.

GLOSSARY

Active oil — An oil which contains chemically active ingredients to promote boundary lubrication.

Annealing — A heating and slow cooling cycle which may remove stress, induce softness, or refine structure.

Boundary lubrication — Lubrication by a solid lubricant (a material which has relatively low shear strength). This lubricant may be introduced in solid form, or it may be formed on fresh metal surfaces by chemical reaction.

Built-up edge (BUE) — A piece of work material which has been strain hardened and pressure welded to the cutting edge of a tool.

Burr — A turned over edge of metal resulting from certain machining operations.

Chatter marks — Surface imperfections on the work surface usually caused by vibrations of the tool and/or workpiece.

Chemical coolant — A cutting fluid which does not contain any mineral oil; usually a true solution in water or a fine colloidal solution.

Chlorine — A common extreme pressure (EP) additive used to promote lubrication.

Coolant — Liquid used to cool the work and tool and to prevent rusting or corrosion; cutting or grinding fluid.

Coupling agent — A mutual solvent, an emulsifier.

Creaming — A concentration of oil droplets of an emulsion near the surface when the emulsion stands quiescent for a sufficient period of time.

Creep-feed grinding — A grinding process in which large volumes of material are removed in a single pass.

Cutting fluid — Fluid (liquid, gas, or mist) applied to the working part of a tool or cutter to promote more efficient machining; coolant; machining lubricant.

Cutting rate — The amount of material removed in a machining operation per unit of time.

Dermatitis	An unnatural condition of the skin.
Ductile	Capable of being deformed when cold.
Emulsifiable oil	A straight oil or blend which contains an emulsifier or coupling agent so it will form a stable emulsion in water.
Emulsifier	A material containing two types of molecular groups, one of which will orient in water and the other in oil. It will, thus, tie together two dissimilar liquids.
Emulsion	An oily mass in suspension in a watery liquid or vice versa.
EP additive	Extreme pressure additive.
Ester	A compound which may be formed by replacement of the acid hydrogen of an acid by a hydrocarbon radical.
Ethane	A gaseous paraffinic hydrocarbon, (CH_3CH_3), occurring in natural gas.
Extreme pressure additive	A compound which reacts with the surface of the metal (or tool) forming thin films of metallic compounds (usually, a chloride, sulfide, or phosphate) that have relatively low-shear-strength.
Extreme pressure lubrication	See "Boundary lubrication." In addition, withstands much higher pressures and temperatures than boundary lubricated surface.
Fatty acid	Any of the series of saturated or unsaturated acids ($C_nH_{2n}O_2$) such as stearic, oleic, and palmitic acids which occur in natural fats and natural oils.
Feed	(Milling) The maximum thickness of material removed per tooth.
Feed	(Turning) The amount of horizontal movement of the tool per revolution of the workpiece.
Feed lines	Spiral pattern produced on work in machining.
Finish	Surface quality or appearance.
Finishing	The final cuts taken to obtain the accuracy and finish.
Halogen	The group of elements: chlorine, fluorine, bromine, and iodine.
Hexane	Any of five volatile liquid hydrocarbons, C_6H_{14}, of the paraffin series.

Hydrodynamic lubrication	Lubrication where the viscosity of the lubricant keeps the surfaces separated by a fluid film.
Inverted emulsion	A dispersion of droplets of water in oil produced when a small quantity of water is mixed with a relatively large quantity of oil.
Machinability	The relative difficulty of a machining operation with regard to tool life, surface finish, and power consumption.
Metallic soap	The reaction product produced when a fatty acid reacts with metal.
Micelle	An aggregation of surface active molecules in a solution.
Microbicide	Any agent which destroys germs or micro-organisms.
Mineral oil	Any oil of mineral origin such as petroleum.
Miscible	Capable of being mixed.
Napthene	Any of a series of saturated cyclic hydrocarbons of the general formula C_nH_{2n}; applied especially to those members occurring in certain kinds of petroleum.
Olefin	Any open-chain hydrocarbon having one or more double bonds.
Paraffin	Any hydrocarbon of the methane series, especially any of the solid members boiling above 572°F (300°C).
Peripheral speed	The speed of any point on the surface of the work (cutter for rotary tool).
RPM	Revolutions per minute.
Soluble oil	See "Emulsifiable oil."
Stearate	A salt or ester of stearic acid.
Stress corrosion	Corrosion facilitated by high residual surface stress imposed by machining or grinding operations.
Sulfo-chlorinated oil	Cutting oil containing sulphur and chlorine.
Sulfur	A common extreme pressure (EP) additive used to promote boundary lubrication.
Superabrasive	An extremely hard abrasive such as CBN or diamond, which is used to manufacture grinding wheels and cutting tools.

Surface active agent	Materials capable of lowering surface and interfacial tensions. See "Wettability."
Synthetic fluids	Products which do not contain any mineral oil and usually form a true solution in water.
Tramp oil	Leakage into the cutting fluid system from hydraulic or lubrication systems of machine tools.
Wettability	The relative ease with which a liquid spreads over a surface.
Wetting agent	An additive which reduces surface and interfacial tension and, thus, facilitates spreading of a fluid over a surface.
Workpiece	The part being machined.
Work hardening	A hardening process which may occur during cold working or machining; strain hardening.

BIBLIOGRAPHY

Albrecht, A. B. "How to Secure Desired Surface Finish in Turning Operations,"*American Machinist, 100* (October 8, 1956), 133–36.
Alcoa Aluminum Handbook. (Pittsburgh, Pennsylvania: Aluminum Company of America, 1962).
Backer, W. R., and E. J. Krabacher. *ASME Transactions, 78* (1956), 1497.
Barker, G. E. "Place of Coolants in Solving Grinding Problems," *Grinding and Finishing, 11*, No. 2, (February 2, 1965), 34–7.
Bartell, F. E., J. L. Culbertson, and M. A. Miller. "Alteration of the Free Surface Energy of Solids," *Journal of Physical Chemistry, 40,* Parts 1, 2, and 3, (October, 1936), 881–904.
Bastian, E. L. H., I. Rozalsky, and K. F. Schiermeter. "New Developments in Metal Working," *Lubrication Engineering, 17*, No. 1, (January, 1961), 40–7.
Bennett, E. O. "Control of Bacterial Spoilage of Emulsion Oils," *Soap and Chemical Specialties* (October and November, 1956).
_____, C. L. Adamson, and V. E. Feisal. "Factors Involved in the Control of Microbial Deterioration – I. Variation in Sensitivity of Different Strains of the Same Species," *Applied Microbiology, 8*, No. 6, (November, 1959).
_____, and R. H. Bauerle. "The Sensitivities of Mixed Populations of Bacteria to Inhibitors," *Australian Journal of Biological Sciences, 13*, No. 2, (1960), 142–49.
_____. "The Role of Sulfate-Reducing Bacteria in the Deterioration of Cutting Emulsions," *Lubrication Engineering,* (April, 1957).
_____. "Factors Involved in the Preservation of Metal-Cutting Emulsions," *Developments in Industrial Microbiology, 3;* 273–85.
_____, G. J. Guynes, and D. L. Isenberg. "The Sensitivity of Sulfate-Reducing Bacteria to Antibacterial Agents (Phenolic Compounds)," *Producers Monthly, 23*, No. 1, (1958) 15–19.
_____, _____, and _____. "The Sensitivity of Sulfate-Reducing Bacteria to Antibacterial Agents – III," *Producers Monthly, 24*, No. 5, (March, 1960) 26–7.
Blake, K. R. "Dynatomics – A New Concept in Metal Removal,"*ASTE Paper No. 23-1,* (March, 1952).
Borsoff, V. N., and C. D. Wagner. "Studies of Formation and Behavior of an Extreme-Pressure Film," *Lubrication Engineering, 13* (February, 1957), 91–99.
Bowden, F. P., and K. E. W. Ridler. "The Surface Temperature of Sliding Metals," "The Temperature of Lubricated Surfaces," *Proceedings of the Royal Society, 154 (Series A),* (London: 1935), 640.
_____ and D. Tabor. *The Friction and Lubrication of Solids* (Oxford Clarendon Press, 1950), 35–7.
_____ and _____. "Mechanism of Friction and Lubrication in Metal Working," *Institute of Petroleum Journal, 40* (1954), 243.
Carlson, V., and E. O. Bennett. "The Relationship Between the Oil-Water Ratio and the Effectiveness of Inhibitors in Oil Soluble Emulsions," *Lubrication Engineering,* (December, 1960).
Carslow, H. S., and J. C. Jaeger. *Conduction of Heat,* (Oxford Clarendon Press, 1948).
Carter, W. A. "Metal Machining," Part 6, "Cutting Fluids," *Machinery, 28* (January 21, 1956), 69–75.
Chao, B. T., L. H. Li, and K. J. Trigger. "The Influence of Tool Geometry on Interface Temperatures," *Tech. Report ORD-1980-4,* (Urbana, Illinois: University of Illinois Department of Mechanical Engineering, April, 1952).
_____, and K. J. Trigger. "Temperature Distribution at the Tool Chip Interface in Metal Cutting," *ASME Transactions, 78* (October, 1955), 1107–21.
Christopherson, D. G., P. L. B. Oxley, and W. B. Palmer. "Orthogonal Cutting of a Work-Hardening Material," *Engineering 186* (London: 1958), 113.
Colding, B., and L. G. Erwell. "Wear Studies of Irradiated Carbide Cutting Tools," *Nucleonics, 11* (February, 1953), 46–9.

Colwell, L. V. "A Method for Studying the Behavior of Cutting Fluids in Wear of Tool Materials," *ASME Paper 57-A-160*, (December, 1957).

_____, and H. Branders. "Behavior of Cutting Fluids in Reaming Steels," *ASME Paper No. 57-A-168*, (December, 1957), 1–5.

Cook, N. H. "Cutting Tool Temperatures," *ASTE Paper No. 21* (1957).

Copper and Copper Alloys, Publication B-32, (Anaconda American Brass Company, Waterbury, Connecticut, 1965).

Cosgrove, S. L., and R. W. Greenlee. "Present Knowledge of Cutting Fluids," *ASTE Report No. 11*, Paper 79 (May 1, 1958), 1–19.

Creveling, J. H., T. F. Jordan, and E. G. Thomsen. "Some Studies of Angle Relationships in Metal Cutting," *ASME Transactions, 79* (January, 1957), 127–38.

"Cutting Fluids Poise for New Sales Splash," *Chemical Week, 81* (October 19, 1957), 91–2; 94–6.

Daasch, F. J., et al., "Evaluating Cutting Fluids in Accelerated Machining Tests," *Lubrication Engineering, 13* (September 9, 1957), 516–20.

Epifanov, G. I., P. A. Rebinder, and L. A. Shreiner. "The Influences of the Nature of a Metal on the Facilitation of Cutting by Adsorption," Doklady Akademii Nauk, CCCR, 66, No. 5 (1949), 879–80.

_____, N. A. Pleteneva, and P. A. Rebinder. "Mechanism of Action of Active Media During Machining of Metals," Doklady Akademii Nauk, CCCR, 97, No. 2 (1954), 277–79.

Ernst, Hans. *Annals of the New York Academy of Science, 53* (1951), 936.

_____ and M. E. Merchant. *Proceedings of Massachusetts Institute of Technology*, Summer Conference on Friction and Surface Finish, 76 (1940).

Feisal, E. V., and E. O. Bennett. "The Effect of Water Hardness on the Growth of *Pseudomonas Aeruginosa* in Metal Cutting Fluids," *The Journal of Applied Bacteriology, 24*, No. 2, (August, 1961), 126–130. 126–130.

Fersing, L. "Carbide High Velocity Turning," *ASME Transactions, 73* (May, 1951), 359–74.

Finnie, I., and M. C. Shaw. "The Friction Process in Metal Cutting," *ASME Paper 54-A-108*, (November 18, 1954).

_____, and E. Rabinowicz. "A Radioactive Study of the Metal-Cutting Process," *Lubrication Engineering, 12* (January and February, 1956), 29–31.

Fisher, R. C. "How Wet Is Your Wet Grinding," *American Machinist/Metalworking Manufacturing, 107*, No. 7, (April 1, 1963), 114–15.

Flemming, C. D., and R. J. Baker. "Controlling the Spoilage of Water Soluble Cutting Fluids," *Lubrication Engineering*, (September, 1960), 414–19.

Fox, H. W., E. F. Hare, and W. A. Zisman, *Journal of Physical Chemistry, 59* (1955), 1097.

Gideon, D., R. Simon, and H. Grover. "Some Thermal and Physical Aspects of Cutting," *ASTE Research Report No. 76, 10* (May 1, 1958), 1–7.

Guynes, G. J., and E. O. Bennett. "Bacterial Deterioration of Emulsion Oils I. Relationship Between Aerobes and Sulfate-Reducing Bacteria in Deterioration," *Applied Microbiology, 7*, No. 2, (March, 1959).

Hablanian, M. "Temperature Rise of the Workpiece in Metal Cutting," Unpublished Master of Science thesis, Massachusetts Institute of Technology, January, 1957.

Hahn, R. S. "On the Temperature Developed at the Shear Plane in the Metal Cutting Process," *First U. S. National Congress of Applied Mechanics Proceedings* (1951), 661–66.

Halverstadt, R. D. "The Development of a Test for Evaluating Grinding Fluids," *Lubrication Engineering, 17*, No. 3, (March, 1961) 127–33.

_____. "The Development of a Test for Evaluating Grinding Fluids," *ASLE Paper No. 58AM 4B-4*, (April 23, 1958).

Harada, M., and N. Shinozaki. "Effect of Grinding Fluids on Grinding," *International Production Engineering Research Conference Proceedings*, (September, 1963), 218–24.

Hays, L. C., and E. J. R. Hudec. "A Tool-Blade Wear Test for Cutting Fluids," *Lubrication Engineering, 10* (February, 1954), 10–23.

Hollander, M. B., and J. E. Englund. "A Thermocouple-Technique Investigation of Temperature Distribution in the Workpiece During Metal Cutting, *ASTE Research Fund Report No. 7*, (1957).

Holmes, P. M. "Development Testing of Cutting Fluids," *Production Engineering, 44*, No. 3, (March, 1965), 130–41.

Husa, H. W., and W. L. Bulkley. "Cutting Fluid Performance," *Lubrication Engineering, 13* (October, 1957), 557–62.

Isenberg, D. L., and E. O. Bennett. "Bacterial Deterioration of Emulsion Oils – II. Nature of the Relationship Between Aerobes and Sulfate-Reducing Bacteria," *Applied Microbiology, 7*, No. 2, (March, 1959).

BIBLIOGRAPHY

Iverson, Warren P. "Direct Evidence for the Cathodic Depolarization Theory of Bacterial Corrosion," *Science, 151,* 986–88.

Kececioglu, D. "Shear Strain Rate in Metal Cutting and Its Effect on Shear Flow Stress,"*ASME Paper 56-A-154,* (1956).

────── and A. Sorensen, Jr. "Coolant Performance Compared," *Tool and Manufacturing Engineer, 45,* No. 5, (November, 1960), 101–6.

──────, "Machining with Single Point Tools," *The Tool Engineer, 8* (January, 1940), 14–16.

Kitzke, E. D., and R. J. McGray. "The Occurrence of Molds in Modern Industrial Cutting Fluids," *Lubrication Engineering,* (March, 1963), pp. 110–13.

Kizilos, A. P. "A Study of the Workpiece Temperature in the Vicinity of the Wear Land in Lathe Turning Operations," Unpublished Master of Science thesis, Massachusetts Institute of Technology, June, 1958.

Kobayashi, S., and E. G. Thomsen. "The Role of Friction in Metal Cutting," *ASME Transactions, 82,* Series B, (1960) 324.

Kohn, E. M. "Device for Studying Cutting Fluid Action at Low Speeds," *Lubrication Engineering, 19,* No. 9, (September, 1963), 371–73.

──────. "Theory on Role of Lubricants in Metal Cutting at Low Speeds and in Boundary Lubrication," *Wear-Usure-Verschleiss, 8,* No. 1 (January and February, 1965), 43–59.

Krabacher. E. J. *Lubrication Engineering, 11* (November, 1955), 397.

Kronenberg, M. "Machining with Single Point Tools," *ASME Transactions, 28* (September, 1940), 725–42; *The Tool Engineer, 8* (January, 1940), 14–16.

Lauterbach, W. E. "Influence of Point of Application of Cutting Fluid on Tool Life," *Lubrication Engineering, 8* (June, 1952), 135–36.

──────, and E. A. Ratzel. "An Investigation of the Flow and Effect of a Cutting Oil in Machining Operations," *Lubrication Engineering, 7* (1951), 15.

Lee, E. H., and B. W. Shaffer. "The Theory of Plasticity Applied to a Problem of Machining,"*ASME Transactions,* (Fluid Applied Mechanics), *73* (1951), A405.

Leone, W. C. "Distribution of Shear-Zone Heat in Metal Cutting," *ASME Transactions, 76* (January, 1954), 121–25.

Letner, H. R. "A Modern Perspective of the Grinding Process," *Grinding and Finishing,* (May, 1955), 36–41.

──────, "Influence of Grinding Fluids upon Residual Stresses in Hardened Steels,"*ASME Paper No. 55-A-123,* (1955).

Liberthson, L. "Bacterial Deterioration of Cutting Oil Emulsions," *Lubrication Engineering,* (December, 1945).

──────. "Effect of Sulphur Bacteria on Corrosion," *Iron and Steel Engineer,* (June, 1947).

Lindert, A. W. "Some Problems Encountered in the Use of Soluble Oils," *Lubrication Engineering,* (October, 1951).

Ling, F. F., and E. Saibel. "On the Tool-Life and Temperature Relationship in Metal Cutting," *ASME Transactions, 78* (1956), 113.

Loewen, E. G. "Thermal Aspects of Metal Cutting," Unpublished Doctor of Science dissertation, Massachusetts Institute of Technology, Cambridge, Massachusetts, 1952.

Manual on Cutting of Metals, American Society of Mechanical Engineers, (New York: 1952).

Mason, J. P., and E. B. Weber. "Foam Cooling Clings," *American Machinist, 98* (April 12, 1954), 174–76.

Merchant, M. E. "Theory of Friction and Its Part in the Metal Cutting Process," Unpublished Doctor of Science thesis, University of Cincinnati, 1940.

──────. "Basic Mechanics of the Metal Cutting Process," *Journal of Applied Mechanics* (September, 1944), pp. A-168–A-175.

──────. "Metal Cutting Research—Theory and Application," *Machining Theory and Practice* (Cleveland, Ohio: ASM, 1950), pp. 5–44.

──────. "Fundamentals of Cutting Fluid Action,"*Lubrication Engineering, 6* (August, 1950), 163–67; 181.

──────. "The Action of Cutting Fluids in Machining," *Iron and Steel Engineer, 27* (November, 1950), 101–8.

──────, and E. H. Ernst. "Principles of Metal Cutting and Machinability," *Tool Engineers Handbook, ASTME,* (New York: McGraw-Hill Book Co., 1945), Section 17, pp. 302–56.

──────, and Norman Zlatin. "New Methods of Analysis of Machining Processes," *Society for Experimental Stress Analysis Proceedings, 3,* No. 2, (1946), 4–27.

──────, ──────, and E. J. Krabacher. "Radioactive Cutting Tools for Rapid Tool-Life Testing," *ASME Transactions, 75* (May, 1953), 549–59.

──────, ──────, and ──────. "Industry Fights Wear with Isotopes," *Nucleonics, 14,* No. 5, (May, 1956), 54.

———, ———, and ———. "Radioactive Cutting Tools for Rapid Tool-Life," *Nucleonics, 14*, No. 5 (May, 1956), 55–7.

Morton, L. S., and A. L. H. Perry. *Engineering, 192*, D. 1, (1961), 716–17.

Muenger, J. R., and N. C. Derby. "Two Measurement Techniques Used in Evaluation of Cutting Fluids," *ASLE Transactions, 3*, No. 1, (April, 1960), 55–60.

Myler, A. B. "Test for Soluble Oils," *American Machinist* (June 2, 1958).

Nakayama, K. "Temperature Rise of Workpiece During Metal Cutting," *Bulletin of Faculty of Engineering, 5* (Japan: Yokohama National University, March, 1956).

Norris, R. H., and E. J. Lownes. "Fundamentals of Cutting Tool Lubrication," *Science Lubrication, 11*, No. 11, (November, 1959), 26–35.

———, A. R. Eyres, and A. C. Brown. "Some Aspects of Improved Straight-Cutting Oil Developments," *Institution of Mechanical Engineers Proceedings*, Lubrication and Wear Conference, (May, 1963), 347–54.

Okoshi, M., and T. Sata. "Friction on the Relief Face of Cutting Tool," *Scientific Papers, Institute of Physical and Chemical Research, 52*, No. 1493, (Japan: Tokyo, 1958), 216.

Opitz, H. "Temperature Field of a Turning Tool and the Reactions in the Zone of Contact," *Microtecnic, 8*, No. 4, (1954), 183–90.

———, and O. Hake. "Wear Analysis of Hard Metal Turning Tools by Means of Radioisotopes," *Microtecnic, 10*, No. 1, (1956), 5–9.

Oxley, P. L. B. "An Analysis for Orthogonal Cutting with Restricted Toolchip Contact," *International Journal of Mechanical Sciences, 4* (1962), 129.

———. "Mechanics of Metal Cutting for Material of Variable Flow Stress," *ASME Paper 62-WA-74* (November, 1962), 1–7.

Paschkis, V. "Temperature Distribution in the Workpiece by Means of Electric Analogy," *ASTE Research Fund Report No. 1*, (1954).

Pekelharing, A. J., and R. A. Schmermann. "Wear of Carbide Tools – Its Effect on Surface Finish and Dimensional Accuracy," *The Tool Engineer, 38* (October, 1953), 51–7.

Petrozzi, E. "Absolute Measurement of Tool Wear, *Nucleonics, 14* (November, 1956), 121.

Petrys, Tibor. "The Action of Cutting Fluids," *Przeglad Mechaniczny, 15*, No. 12, (1965), 458–9.

———. "Investigation of Physical and Chemical Nature of Performance of Fluids During Machining Process," *Mechanik, 31*, Nos. 8 and 9, (August-September, 1958), 397–9.

Pleteneva, N. A., and P. A. Rebinder. "A Physico-Chemical Investigation of the Cooling Properties of Liquids Used During Machining of Metals," Isvestiya Akademii Nauk, CCCR, Technical Series Section, *12*, (1946), 1823–29.

——— and ———. "Effect of the Surface Activity of the Liquid Medium on Processes of Cutting and Machining of Metals," Doklady Akademii Nauk, CCCR, *62*, No. 4, (1948), 501–504.

——— and G. I. Epifanov. "Relationship of Liquid Media During Free Planing and the Physico-Chemical Nature of the Medium and the Metal," Doklady Akademii Nauk, CCCR, 77, No. 6, (1951), 1051–53.

Pomey, J., C. M. Prevost, and P. Quantin. "Cutting Fluids for Tapping, Testing Method," *CIRP Annalen, 11*, No. 2, (1962), 104–10.

"Properties and Selection of Metals," *Metals Handbook* (8th ed.; Vol. I, Ohio: American Society of Metals, 1964).

Proskuryakov, Yu. G., N. F. Belov, and V. N. Petrov. "Use of Mist Cooling for Extending Cutting Tool Life," *Stanki i Instrument, 32* (June, 1961), 25–9. (See also English translation in *Machines and Tooling, 32*, No. 6, (1961), 29–33.

Rebinder, P. A. *Nature*, 159 (1947), 866.

Reichenbach, G. S. "Experimental Measurement of Metal-Cutting Temperature Distributions," *ASME Paper 57-SA-53*, (1957).

Rossmoore, H. W. "Correlation of Coliform Activity and Anaerobic Sulfate Reduction with Deterioration of Cutting Fluids," *Lubrication Engineering*, (May, 1962).

Roubik, J. R. "Carbide Steel Milling with Cutting Fluids – A Progress Report," *Lubrication Engineering, 8* (October, 1952), 235–37; 261.

Sato, T., and M. Mizuna. "The Friction Process on Cutting Tool and Cutting Mechanism," *Journal of the Scientific Research Institute, 49*, No. 1, (1955), 163–74.

———. "Features and Principles of Through Wheel Coolant Grinding," *Faculty Engineering Technology Reports, 28*, No. 1, (Tohoku University: 1963), 87–102.

Schallbroch, H., and H. Schaumann. "Die Schnit-temperature beim Drehvorgang und ihre Anwendung als Zerspanbarkeitskennziffer," *VDI Zeit, 81* (March 13, 1937), 325–30.

Schepers, A., and Evon Blumenstein, "Einfluss der Schneidocltemperatur auf die Streuungen von Einzelwerten beil Standzeitversuchen," *Werkstattstechnik, 50*, No. 6, (June, 1960), 310–12.

Schmidt, A. O. et al, "Ceramic and Carbide Tool Performance Tests, Part II," *ASME Paper No. 57-SA-101*, abstracted in *Mechanical Engineering, 79* (October, 1957), 967.

Shaw, M. C. "Cutting Fluid Theory" in *Machining Theory and Practice*, (Ohio: American Society for Metals, 1950) 45–68.

_____. *Metal Progress, 42* (1942), 85.

_____. "Mechanical Activation—Newly Developed Chemical Process," *ASME Transactions, 70* (1948).

_____. "On the Action of Metal-Cutting Fluids at Low Speeds," *Wear-Usure-Verschleiss, 2*, No. 2, (February, 1959), 217–27.

_____. "Mechanical Activation, a Newly Developed Chemical Process," *Journal of Applied Mechanics, 15* (March, 1948), 37–44.

_____. "Machining Costs and Their Influences on Metal Cutting Practice," *Microtecnic, 13*, No. 3.

_____., and P. A. Smith. "Evaluation of Water-Base Cutting Fluids," *Report to Watertown Arsenal*, (1956).

_____., N. H. Cook, and I. Finnie. "The Shear-Angle Relationship in Metal Cutting," *ASME Transactions, 75* (February, 1953), 273–88.

Sherwood, P. W. "What to Look For in Chemical Cutting and Grinding Fluids," *Scientific Lubrication, 12*, No. 9, (September, 1960), 25–6.

Shono zaki, Tomonobu. "Friction in the Cutting of Metals as Related to Cutting Oils," *Yukagaku*, Kanagawa University, Yokohama, *10*, (1961), 71–6.

Sluhan, C. A. "Cutting Fluids," *ASTME Paper No. 399, 61*, Book I (1962).

_____. "Some Considerations in the Selection and Use of Water Soluble Cutting and Grinding Fluids," *Lubrication Engineering 16*, (1960), 110–18.

Sniechowski, R. "Study of Oil Emulsions as Coolants," *Mechanik, 31*, Nos. 8 and 9, (August and September, 1958), 387–92.

"Steel Specifications and Standard Compositions," *SAE Handbook*, (New York: 1953).

Sudholz, L. H. "The Evaluation of Cutting Fluids in Laboratory Equipment," *Lubrication Engineering, 4* (January, 1948), 18–24.

Takeyama, H., and E. Usui. "The Effect of Tool-Chip Contact Area in Metal Machining," *ASME Paper No. 57-A-45*; abstracted in *Mechanical Engineering, 80* (February, 1958).

Tant, C. O., and E. O. Bennett. "The Isolation of Pathogenic Bacteria from Used Emulsion Oils, *Applied Microbiology, 4*, No. 6, (November, 1956), 332–38.

Trigger, K. J., and B. T. Chao. "The Mechanism of Crater Wear of Cutting Tools," *M. E. Technical Report Ord. 1121-2* University of Illinois, (August, 1955).

_____ and _____. "The Mechanism of Crater Wear of Cemented Carbide Tools," *ASME Transactions, 78* (1956), 119–25.

Vintner, F., and J. Kolar. "Influence of Cutting Fluids on Economy of Machining," *Czechoslovak Heavy Industry, 7* (1964), 35–42; see also abstract in *Engineers' Digest, 35*, No. 9, (September, 1964), 103–5.

Wallace, W. P. "Tool Forces and Tool-Chip Friction in Orthogonal Machining," Unpublished Fellowship thesis, Royal College of Advanced Technology, Salford, 1962.

_____, and G. Boothroyd. "Tool Forces and Tool-Chip Friction in Orthogonal Machining," *F. Mech. Engng. Science, 6* (1964), 74.

Weiner, J. H. "Shear-Plane Temperature Distribution in Orthogonal Cutting," *ASME Transactions, 77* (November, 1955), 1331–41.

Wheeler, H. O., _____. "Bacterial Inhibitors for Cutting Oil," *Applied Microbiology, 4*, No. 3, (May, 1956), 122–26.

Wilcox, A. "Spraying of Coolants Increases Tool Life," *Tool Engineer, 40*, No. 4, (April, 1958), 111–12.

Wolfe, K. J. B., M. D. Kinman, and G. Lennard. "A Preliminary Investigation of the Effectiveness of Chlorinated Hydrocarbon Compounds as Cutting Oil Additives," *Institute of Petroleum Journal, 40* (September, 1954), 253–56.

Yakunin, G. K. "Increasing Tool Life of High-Speed Steel Tools During Cutting with a Supply of Gaseous Oxygen in the Chip-Formation Zone," *Stanki i Instrument, 26* (April, 1955), 21

Yanis, R. J., and G. F. Wolfe. "Test Procedures for Evaluation of Cutting Fluids," *Lubrication Engineering, 16*, No. 4, (April, 1960), 164–70.

Zener, C. *Elasticity and Anelasticity of Metals* (Chicago: University of Chicago Press, 1948), pp. 126–30; 146–50.

Zisman, W. A. "Relation of Equilibrium Contact Angle to Liquid and Solid Constitution," *Contact Angle, Wettability, and Adhesion*, Ch. 1, American Chemical Society (Washington, D. C., 1964).

INDEX

A

Abrasive cutoff, 70, 95
Acceptance tests (*See* Laboratory test procedures)
Adsorptional plasticizing (*See* Hardness reduction)
Aerobic bacteria, 111, 126
"Air as gaseous cutting fluid", 46
Air-carried mist, 56 - 57
    aspirator and pressure-fed generators, 58
    portable systems, 58
Alcohols, poly or complex, 36, 43
Aluminum and alloys, 36, 64 - 70, 110
Amides, 44
    condensates, 36
Amine, 44
    soaps, 43
Amontons' law, 27
Anaerobic bacteria, 45, 111, 126 - 127
Animal oils, 2, 36 - 37, 41
Antilubrication, 30
Application of fluids (*See* fluid application)
Aqueous fluids, 7, 23
Argon, 47, 60 (*table*)
Aspirator-type mist generator, 58
Atmospheric environment, 97

B

Bacteria, 38, 45, 121 - 122, 125 - 127
Bacterial effects and prevention, 126 - 127
Bacterial spoilage, 12
Benzene, 21
Beryllium and alloys, 64 - 71
Borates, 44
Boring, 64, 76
Boundary lubrication, 27 - 29, 36 - 38, 45, 57
Box turning, 64, 78
Brass, 35 - 36, 39, 96, 110
Brittle metals, 9
Brittleness, chip, 22
Broaching, 42, 68, 72, 86
Bromine, 107
Bronze, 39
Built-up edge, 7 - 9, 31, 39

C

Capillary action of fluids, 18
Carbon dioxide, 18, 31 - 32, 47, (*table*), 60

Carbon tetrachloride, 15, 19, 21, 24, 29, 32, 47, 115
Cast iron, 64 - 71, 96, 103, 110
Castor oil, 38
Caustic embrittlement, 110
CBN grinding, 183 - 184
Central systems, 127, 130
Centrifugation, 127 - 128, 130 - 131, 133
Centrifuging, 131
Ceramic tools, 73
Chasing, 68, 87
Chatter, 164
Chemical action, 14
Chemical additives, 110
    bromine, 107
    chlorine, 2, 4, 11, 14, 27 - 30, 35 - 41, 43, 45, 47, 73, 107, 112
    fluorine, 107
    iodine, 107, 109 - 110, 115
    phosphorus, 2, 4, 35, 38 - 41, 43, 45, 107, 109, 133
    sulfur, 2, 4, 11, 14, 21, 23, 27 - 30, 35 - 41, 43, 45, 73, 107
Chemical compatibility, 107
Chemical fluids, 1, 44 - 45, 120
    solutions, 23
    water miscible, 43
        true solution type, 44
        wetting agent type, 44
        wetting agent type with EP lubricant, 44
Chemically active fluids, 11, 25
Chilled fluid application, (*See* Refrigerated fluids)
Chip area compression, 22
Chip hardness, 8, 22
Chip shortening ratio, 6
Chip thickness, 6 - 7, 9 - 10, 39
Chlorine additives, 2, 4, 11 - 12, 14, 28 - 30, 35 - 36, 38, 40 - 41, 43, 45, 47, 73, 107, 110, 112 - 113
Cleaning machine tools and circulation systems, 132
Cobalt base alloys, 64 - 71
Coconut oil, 38
Coefficient of friction, 7, 27
Cold test, 119
Colloidally dispersed solid lubricants, 28
Compounded cutting oils, 40
Compressed air, 24, 46
Compressive stress, 29
Cooling mechanism, 12, 19, 21, 23, 25
    aqueous fluids, 22
    benzene, 21

carbon tetrachloride, 21
chemical solutions, 23
chip area compression, 22
chip brittleness, 22
chip hardness, 22
cutting oils, 22
cutting ratio, 22
cutting temperature, 20
dry cutting, 21 - 22
internally-cooled cutting tool, 19, 21
oil in water emulsion, 21, 23
shear zone, 21
sulfurized oil, 21, 23
surface finish, 22
temperature effects of cutting fluids, 21, (table) 23
thermocouple measurement of temperature, 20 - 21
water as cutting fluid, 21
water surfactant solution, 21
wear effects, 20
Copper and alloys, 39, 64 - 71
Corrosion, 12, 46, 102 - 104, 107 - 109, 121 - 122
of machines, 110
prevention, 12, 37, 40, 42, 44, 46, 126
Counterboring, 68, 85
Countersinking, 68, 86
Couplers, 43
Crater wear, 9
Creep-feed grinding, 181 - 184
Critical state of fluids, 13
Cutoff, 64, 72, 78
abrasive, 70, 95
Cutting at higher speeds, 8
built-up edge, 8
chip and tool seizure, 9
crater wear, 9
cutting ratio, 8
cutting temperature vs. cutting forces, 9
deformation zones, 8
plastic deformation, 8
surface finish, 8
tool wear, 9
Cutting at low speeds, 8
built-up edge, 8
secondary deformation zone, 8
surface finish, 8
vibration harmonics, 8
work-hardening, 8
Cutting forces, 9, 11, 19, 21, 25, 29
normal, 25
tangential, 25
Cutting hard and brittle metals, 9
built-up edge, 9
chemical reactions, 9 - 10
chip hardness, 9
chip thickness, 9

cutting forces, 9
dry cutting, 9
effect of hardness on fluid selection, 10
plastic flow, 9
temperature rise, 9 - 10
tool wear, 10
Cutting oils, 2, 22, 24, 36 - 37, 39, 42, 107, 119
Cutting ratio, 6, 8, 22
Cutting temperature, 20, 25
Cutting tools, 72

D

Deformation zone, 30
primary, 7
secondary, 7 - 8
Deionization, 123
Dermatitis, 46 - 47
Diamond grinding, 184 - 185
Dimensional stability, 12
Disposal, 119, 189 - 190
disposal of water-base fluids, 133
RCRA, 187, 189
Drilling, 50, 54, 66, 84
Dropout rate, 106
Dry cutting, 8 - 9, 21 - 23, 25
Dry grinding, 11
Dry refrigerated cutting, 25

E

Economics of fluid selection, 143 - 144, 146, 148, 150, 152, 154
Electrically-assisted (EAG) grinding, 55
Elemental diffusion, 110
Embrittlement effect, 29 - 30, 110
Emulsifiable cutting fluids (See Emulsions)
Emulsifiable oils, extreme pressure, 43
Emulsifiers, 42
Emulsions, 2, 41 - 42, 111, 119 - 120
mineral oils, emulsified, 43
oil in water, 21, 23, 35
oil, water miscible, 41 - 43, 45, 59
highly-fatted, 43
EP (See Extreme pressure)
Esters, 2, 36, 41, 43
Extreme pressure additives, 37, 43, 57
Extreme pressure emulsions, 41
Extreme pressure lubricants, 36 - 40
Extreme pressure semichemical fluids, 2

F

Facing, 64, 77
Fatty acids, 27, 38, 43
Fatty amine condensates, 36

INDEX

Fatty oils, 4, 28, 35 - 36, 38, 43
Ferric chloride, 29
Filtration, 131
Fire hazards, 12, 57, 138
Flood application of fluids, 52 - 53, 55, 59
    passing fluid through the wheel, 55
    special nozzles for grinding, 55
Flowstress curve, 28
Fluid action in cutting, 12 - 13, 15, 17
    (central, individual), 120, 127
    bacterial spoilage, 12
    centrifuging, 131
    cleaning methods, 129, 131
    cooling mechanism, 12
    corrosion prevention, 12
    corrosivity and toxicity, 12
    cutting forces, 12
    dimensional stability, 12
    disposal of water-based fluids, 133
    Filtration, 131
    fire hazards, 12
    flushing of chips, 12
    handling methods, 119
    lubricating mechanisms, 12
    lubrication of machine tool slideways, 12, 140
    machinability, 12
    magnetic separators, 131
    physical instability, 12
    product temperature, 12
    quality control, 119 - 120, 122, 124, 126, 128, 130, 132, 134, 136
    seizure between chip and tool, 12
    settling, 130
    shear flow characteristics, 12
    storage, 119
    straining, 130
    surface finish, 12
    systems (central, individual), 120, 127
    tool wear, 12
Fluid application
    air-carried mist, 56 - 57
    chilled fluids, 59, 61
    compressed gases, 60
    flood application, 52 - 53, 55
    high velocity, 11
    manual, 50 - 51
Fluid film, 26 - 28
    lubricants, 37
Fluid function in cutting, 7
Fluid function in grinding, 10 - 11
Fluid selection, 64 - 69, (tables), 71
    economic considerations, 143 - 144, 146, 148, 150, 152, 154
Fluid states and properties, 13
    chemical action, 14
    critical, 13

gas diffusion, 13
gaseous, 13
interfacial tension, 14, 36
liquid, 13
shear properties of metal, 14
thermal properties, 13 - 14
viscous resistance, 13
wetting ability, 13
Fluid types
    air, 46
    argon, 47
    carbon tetrachloride, 47
    chemical fluids, water miscible, 43, 45
    chemically active, 9, 11
    cutting oils, 36 - 37, 39
        emulsified mineral oils, 36
        emulsified oils, water miscible, 41
        extreme pressure emulsifiable oils, 43
        highly-fatted water miscible fluids, 43
        mineral oils, 43
    gaseous fluids, 46
        air, 46
        argon, 47
        helium, 47
        nitrogen, 47
    semichemical fluids, water miscible, 43, 45
    trichloromethane, 47
Fluids as coolants, 5, 7, 9, 11, 19, 21, 23 - 25, 29, 38, 43, 45
    at higher speeds, 8
    cooling mechanism, 26 - 27
    in cutting hard and brittle metals, 9
    in grinding, 10, 38
Fluids as lubricants, 5, 8, 22, 24, 26 - 27, 29, 46
    at higher speeds, 8
    at low speeds, 8
    in grinding, 10
    lubrication mechanism, 26 - 27
Fluorine, 107
Flushing of chips, 12
Foam depressants, 43
Foaming, 42 - 44, 46, 100, 102, 105
Forming, 64, 78
Four-Ball tester, 28
Freon, 47
Friction, 26
    coefficient of, 6
    primary zone, 7
    sliding, 7, 27
    wear (abrasive), 8
Fungi, 43, 121, 129

G

Gas diffusion, 13
Gaseous states of fluids, 13

Gases, 17, 46
  argon, 47
  carbon dioxide, 47
  freon, 47
  helium, 47
  nitrogen, 47
Gear cutting, 70, 92
  hobbing, 70, 92
  lubricants, 140
  shaping, 70, 93
  shaving, 70, 93
    rack, 93
    rotary, 93
Gear wheel milling, 24
Geometrical relationships, 6 - 7
  built-up edge, 7
  chip thickness, 6
  coefficient of friction, 7
  cutting ratio or "chip shortening ratio", 6
  depth of cut, 6
  orthogonal cutting, 7
  primary deformation zone, 7
  primary friction zone, 7
  rake angle, 7
  secondary deformation zone, 8
  shear angle, 6
  shear plane, 7, 20
  shear stress, 7
  sliding friction, 7
  slip-line theory analysis, 7, 10
Gold alloys, 110
Graphite, 28
Grinding, 10, 55, 68, 88 - 89, 91
  CBN grinding, 184 - 185
  centerless, 68, 90
  chemically active fluids, 11
  chip breaker, 57
  creep-feed grinding, 181 - 184
  crush, 42
  cylindrical, 68, 91
  diamond grinding, 184 - 185
  disc-type surface, 44
  double-disc, 43
  dry grinding, 11
  flat-bed, 43
  fluid wetting characteristics, 11
  forces, normal and tangential, 11
  form, 42
  formed wheel, 70, 91
  gear, 70, 91
  generation, 70
  grinding ratio, 11
  heat generation and removal, 11
  high-velocity fluid application, 13
  internal, 68, 89
  rake angle, 10
  random orientation of cutting faces, 10
  ratio, 11
  residual stress, 11
  speed, 10
  superabrasives, 181 - 182, 184 - 185
  surface, 11 - 12, 49, 68, 89
  temperature, 12
  thread, 42, 70, 91
  tool wear and bond fracture, 11
  wheel life, 11
Grooving, 64, 72, 77
Gundrilling, 43, 54, 66, 85

H

Halogens, 107, 109 - 110
Handling methods for fluids, 119
Hardness
  chip, 22
  effect of on fluid selection, 10
  reduction, 28
  water, 97, 120 - 122, 124
Heat transfer coefficient, 17
Heavyduty solubles, 43
Helium, 47
High-velocity application of gases, 18
Highjet fluid application technique, 18
Honing, 70, 95
Humectants, 44
Hydraulic systems, 138
Hydrocarbon, non-polar, saturated, 37
Hydrodynamic lubricants, 36, 56
Hydrogen embrittlement, 110

I

Individual systems, 120
Interfacial tension, 14, 36, 44
Intergranular corrosion, 107
Intergranular diffusion, 19
Intergranular slip, 30
Internally cooled cutting tools, 21
Iodine, 107, 109 - 110, 115
Iron sulfide, 27 - 28

L

Laboratory test procedures for acceptance, 99
  chemical, 101
    cast iron corrosion, 103
    machinability, 105
    nonferrous corrosion, 102
    pH, 102
    putrefaction, 101
    steel corrosion, 104
    water compatibility, 102
  mechanical, 105

dropout rate, 106
   foaming, 105
   oil reclamation, 106
   tramp oil, 106
  physical, 99
   paint, 101
   residue, 100
   stability, 99
Lapping, 96
Lard oil, 35, 43
Liquid state of fluids, 13
Lubrication mechanism, 12, 43 - 44
  Amontons' law, 27
  boundary lubrication, 27 - 28
  chlorine compounds, 27
  coefficient of friction, 27
  colloidally dispersed solid lubricants, 28
  extreme pressure lubrication, 25
  fatty acids, 27
  fatty oils, 28
  fluid films, 26
  four-ball tester, 28
  friction, 26
  metal salts, 27
  "oiliness", 27
  seizure between workpiece and tool, 27
  sliding friction, 27 - 28
  sulfur compounds, 27
  surface finish, 22, 28
  tool wear, 28
  viscosity, 26
Lubrication of slideways, 12

M

Machinability, 12, 105
Machine corrosion, 110
Machine tool lubricants, 137
Magnesium and alloys, 36, 64 - 71, 110
Magnetic separators, 131
Manual application of fluids, 50 - 51
Marine oils, 36, 38
Metal salts, 27, 41
Metallic soaps, 36, 108
Metallurgical compatibility, 107
Microbicides, 43, 112, 114 - 115, 127, 133
Microcracks, 18, 28 - 30
Milling, 24 - 25, 49, 51 - 52, 58 - 59, 79 - 84
  end, 58, 64, 80
  face, 64, 79
  formed cutters, 64, 82
  hollow milling, 66, 84
  milling saw, 66, 83
  side, 82
  slab (plain), 66, 81
  slot, 66, 82
  thread milling, 66, 84

Mineral oils, 36, 45 - 46
  chemical additives, 38
  compounded cutting oils, 40
  naphthenic, 36
  paraffinic, 36
  polar additives, 36
  solid lubricants, 41
Mist application of fluids, 56 - 57
  portable, 58
Mist cooling, 18
Mist generators
  aspirator, 58
  pressure-fed, 58
Mold, 43
Molybdenum disulfide, 28, 41
Multipurpose machine tool lubricants, 141

N

Naphthenic acids, 43
Naphthenic mineral oils, 36, 43
Nickel base alloys, 64 - 71
Nitrates, 115
Nitrites, 115
Nitrogen, 18, 24, 47, 59, 65 - 71
Normal emulsions, 42, 45 - 46
Nozzle, 49, 52 - 53, 55 - 61, 67
Numerically-controlled machines, 50

O

Oil hole drills, 54
Oiliness, 27 - 28, 35, 46
Oleic acid, 28, 37, 108
Operator protection, 12 - 13, 15, 17, 46 - 47, 57, 60, 98, 111, 113
Organo-metallic film, low-shear, 37
Orthogonal cutting, 6
  slip-line-theory analysis, 7
Oxidation, 36, 47

P

Paint test, 101
Palm oil, 38
Paraffinic mineral oils, 36
Partial boundary lubricants, 36
  alcohols, poly or complex, 36
  condensates, 36
  esters, 36
  fatty acids, 36
  fatty oils, 36
Penetration, 28
Petroleum oil, 2
  naphthenic, 2
  paraffinic, 2

Petroleum sulfonates, 43
pH, 102, 112, 114
Phosphates, 44
Phosphorus additives, 2, 4, 35, 38 - 41, 43, 45, 107, 109, 133
Plastic deformation, 8
Plastic flow, 9
Plasticity, 19, 29 - 30
Polar additives, 36 - 38
  animal origin, 36
  marine, 36
  partial boundary lubricants, 36
  vegetable origin, 36
Polar film, 37
Potassium chloride, 107
Precutting fluid penetration, 19
Pressure-fed type mist-generator, 58
Putrefaction, 12, 101

Q

Quality control of fluids, 119
  bacterial effects and prevention, 125, 127
    central systems, 127
  cleaning machine tools and circulating systems, 132
  disposal of water-base fluids, 133
  fluid cleaning methods, 129, 131
    centrifuging
    filtration, 131
    flotation, 131
    magnetic separators, 131
    settling, 130
    straining, 130
  fungi, 129
  handling methods, 119
    central vs individual systems, 120
    storage, 119
  water sources and composition, 120 - 121, 123

R

Rake angle, 7, 10
RCRA, 187, 189
Reaming, 59, 68, 86
Rebinder effect, 28
Recessing, 72
Reclamation of oil, 106
Recycling, 134, 189
Refractory metals, 64 - 71
Refrigerated fluids, 23, 25
Refrigerated gases, 25, 46, 60
Residue, 100, 111, 121 - 123
Ring distributor, 49, 53
Rosin soaps, 43

S

Sawing, 70, 94
  contour, 60
Season cracking, 110
Secondary fluid penetration, 18
Seizure between chip and tool, 9, 12, 27, 46, 162 - 164
Semichemical fluids, 2, 45
  water miscible, 43, 45
Semisynthetic fluids (*See* Semichemical fluids)
Settling, 130
Shear angle, 6, 29
Shear plane, 7, 29
Shear properties of metal, 14
Shear stress, 7
Shear zone, 21, 39
Shear-strength-reduction mechanism, 28
  "adsorptional plasticizing", 29
  boundary lubrication, 28 - 29
  carbon tetrachloride, 29
  chlorine, 29
  coefficient of friction, 29
  compressive stress, 29
  embrittlement effect, 29
  extreme pressure lubricants, 29
  ferric chloride, 29
  flowstress curve, 28
  hardness reduction, 28
  microcracks, 28 - 29
  "oliness", 28
  oleic acid, 28
  penetration, 28 - 29
  plasticity, 29
  "Rebinder effect", 28
  shear angle, 29
  shear planes, 29
  slip planes, 28 - 29
  strain-hardening, 28 - 29
  stress concentration, 29
  sulfur, 29
  surfactants, 28
  tin, 28
Slip planes, 28 - 30
Soaps, 37 - 38, 42, 108, 131
Sodium chloride, 107
Solid lubricants, 41
  colloidally dispersed, 28
Sperm oil, 38, 43
Spindle and bearing lubricants, 139
Spot facing, 68, 86
Stability test, 100
Staining, 39, 108, 127
Stearic acid, 108
Steels, 35 - 36, 39, 47, 64 - 69, 71, 104, 107, 11
Stoddard solvent, 59 *(table)*

Storage, 119
Strain-hardening, 28 - 30
Straining, 130
Stress concentration, 29
Stress corrosion cracking, 47, 107 - 110
Stress, residual, 7, 11
Sulfo-chlorinated fatty oil, 40, 73
Sulfo-chlorinated mineral oil, 40, 73
Sulfur additives, 2, 4, 11, 14, 23, 27, 29 - 30, 36, 38 - 41, 43, 45, 73, 107
Sulfurized fat, 39
Sulfurized oil, 21, 23, 39
Super-fatted emulsions, 2, 41
Super-fatted semichemical fluids, 2
Superabrasives, 180 - 181, 183 - 184
Surface finish, 8, 11, 20, 22, 28 - 29, 39, 163, 166
   microinch, 166
Surface tension, 17, 42, 44
Surface-active solutions, 1, 44
Surfactants, 28
Synthetic fluids (*See* Chemical fluids)

T

Tapping, 47, 50, 59, 68, 87
Temperature effects on cutting fluid action, 23
   carbon dioxide, 24 - 25
   carbon tetrachloride, 24
   chemically active fluids, 25
   compressed air, 24
   cutting forces, 25
   cutting oil, 24
   cutting temperatures, 25 - 26
   deep hole drilling, 24
   dry cutting, 21 - 22, 24 - 25
   dry refrigerated cutting, 25
   gear wheel milling, 24
   milling, 25
   nitrogen, 24
   oil in water emulsion, 23
   refrigerated fluids, 23, 25
   refrigerated gases, 25
   sulfurized mineral oil, 23
   tool wear, 24 - 25
   trichloroethylene, 25
   turning, 25
   viscosity, 24
Tests, acceptance (*See* Laboratory test procedures)
Thermal properties of fluids, 13 - 14 (*table*)
   aqueous liquids, 14
   gases, 14
   heat of vaporization, 13
   organic liquids, 14
   specific heat, 13
   thermal conductivity, 13
   viscosity, 13
Thermocouple temperature measurement, 20 - 21
Threading, 59, 68, 87
Tin, 28
Titanium and alloys, 64 - 71, 107, 109
Tool wear, 8, 10, 12, 22, 24 - 25, 28, 39, 47, 60, 164
Toxicity, 12, 47, 57, 115
Tramp oil, 106, 131
Transfer machines, 59
Transgranular fracture, 30
Trepanning, 64, 79
Trichloroethylene, 25, 59, 107, 115
Trichloromethane, 47
Troubleshooting fluid applications, 161 - 165, 167 - 178
True solutions, 1
   chemical fluids, 44
Turning, 25, 51 - 52, 64, 76

V

Vegetable oils, 4, 38
Vibrational harmonics, 8
Viscosity, 13, 24, 26, 42

W

Waste minimization, 134
   centrifugation, 127 - 128, 130 - 131, 133
   disposal, 119, 133, 189 - 190
   RCRA, 187, 189
   recycling, 134, 189
Water
   as cutting fluid, 21, 35
   compatibility, 102
   hardness, 97, 120 - 122
   miscible fluids (*See* Emulsions)
   soluble oils (*See* Emulsions)
   sources and composition, 120 - 121, 123
   surfactant solution, 21
Way lubricants, 12, 140
Wear effects, 20
Wetting ability, 13, 47
Wetting action, 4, 11
Wetting agents, 42
   chemical fluid, 44
   nonionic, 43
Wetting and penetration, 17, 36, 44
   exposed surfaces, 18
   gases directed at high speed, 17
   "high-jet" technique, 18

intergranular diffusion, 19
  mist cooling, 18
  precutting penetration, 19
  secondary penetration phase (capillary
     action), 18 - 19
  small clearances, 18
  surface tension, 18
Wheel life, 11
Work hardening, 9

Zeolite softner, 121 - 122